BIOLOGY LABORATORY MANUAL

FOR THE TELECOURSE

CYCLES OF LIFE: EXPLORING BIOLOGY

Third Edition

Jerri K. Lindsey
Carolyn C. Robertson
Terri J. Lindsey

Tarrant County Junior College

Wadsworth Publishing Company
I(T)P® An International Thomson Publishing Company

Belmont, CA • Albany, NY • Bonn • Boston • Cincinnati • Detroit • Johannesburg
London • Madrid • Melbourne • Mexico City • New York • Paris
Singapore • Tokyo • Toronto • Washington

COPYRIGHT © 1997 by Wadsworth Publishing Company
A Division of International Thomson Publishing Inc.
I(T)P ® The ITP logo is a registered trademark under license.

Printed in the United States of America
1 2 3 4 5 6 7 8 9 10

For more information, contact Wadsworth Publishing Company, 10 Davis Drive, Belmont, CA 94002, or electronically at http://www.thomson.com/wadsworth.html

International Thomson Publishing Europe
Berkshire House 168-173
High Holborn
London, WC1V 7AA, England

International Thomson Editores
Campos Eliseos 385, Piso 7
Col. Polanco
11560 México D.F. México

Thomas Nelson Australia
102 Dodds Street
South Melbourne 3205
Victoria, Australia

International Thomson Publishing Asia
221 Henderson Road
#05-10 Henderson Building
Singapore 0315

Nelson Canada
1120 Birchmount Road
Scarborough, Ontario
Canada M1K 5G4

International Thomson Publishing Japan
Hirakawacho Kyowa Building, 3F
2-2-1 Hirakawacho
Chiyoda-ku, Tokyo 102, Japan

International Thomson Publishing GmbH
Königswinterer Strasse 418
53227 Bonn, Germany

International Thomson Publishing Southern Africa
Building 18, Constantia Park
240 Old Pretoria Road
Halfway House, 1685 South Africa

All rights reserved. No part of this work covered by the copyright hereon may be reproduced or used in any form or by any means—graphic, electronic, or mechanical, including photocopying, recording, taping, or information storage and retrieval systems—without the written permission of the publisher.

ISBN 0-534-50459-0

TABLE OF CONTENTS

	Telecourse Lesson and Corresponding Laboratory Exercise	iv
	Laboratory Kit Supplies	v
	Introduction	vi
	A Message to the Instructor	vii
	A Message to the Student	viii
1	The Scientific Method	1
2	Basic Chemistry	15
3	The Cell and Cell Processes	21
4	Photosynthesis	31
5	Mitosis and Meiosis	39
6	Introduction to Genetics	49
7	Human Genetics	61
8	DNA Replication	69
9	RNA and Protein Synthesis	77
10	Evolution	85
11	Classification	95
12	Viruses, Bacteria, and Protistans	103
13	Fungi and Plants	113
14	Survey of the Animal Kingdom	121
15	Plants: Tissues, Nutrition, and Transport	141
16	Plants: Reproduction and Development	151
17	Animal Tissues	161
18	Biomechanics	173
19	The Cardiovascular System	189
20	Immunity	203
21	Parasitism	209
22	The Respiratory System	219
23	The Digestive System	235
24	Nutrition	249
25	The Excretory System	263
26	The Nervous System	271
27	Sensory Perception	285
28	Reproduction	303
29	Embryology	315
30	Ecology: Populations and Communities	323
31	Ecology: Ecosystems and the Biosphere	331
32	Ecology: The Human Factor	339
33	Behavior	347
34	A Field Trip	355

TELECOURSE LESSON
AND CORRESPONDING LABORATORY EXERCISE

Telecourse Lesson	Laboratory Exercise
Lesson 1—Biological Concepts	Scientific Method
Lesson 2—Chemical Foundations	Basic Chemistry
Lesson 3—Cell Structure and Function	Cell and Cell Processes
Lesson 4—Metabolism	
Lesson 5—Energy In/Energy Out	Photosynthesis
Lesson 6—Mitosis and Meiosis	Mitosis and Meiosis
Lesson 7—Patterns of Inheritance	Introduction to Genetics
	Human Genetics
Lesson 8—DNA Structure and Function	DNA Replication
Lesson 9—Proteins	RNA and Protein Synthesis
Lesson 10—Microevolution	Evolution
Lesson 11—Macroevolution	Classification
Lesson 12—Viruses, Bacteria, and Protistans	Viruses, Bacteria, and Protistans
Lesson 13—Fungi, Plants, and Animals	Fungi and Plants
	Survey of the Animal Kingdom
Lesson 14—Plants: Tissues, Nutrition, and Transport	Plants: Tissues, Nutrition, and Transport
Lesson 15—Plants: Reproduction, and Development	Plants: Reproduction and Development
Lesson 16—Animals: Structure and Movement	Animal Tissues
	Biomechanics
Lesson 17—Animals: Circulation	The Cardiovascular System
Lesson 18—Animals: Immunity	Immunity
	Parasitism
Lesson 19—Animals: Respiration	The Respiratory System
Lesson 20—Animals: Digestion and Fluid Balance	The Digestive System
	Nutrition
	The Excretory System
Lesson 21—Animals: Neural Connection	The Nervous System
	Sensory Perception
Lesson 22—Animals: Endocrine Control	
Lesson 23—Animals: Reproduction and Development	Reproduction
	Embryology
Lesson 24—Populations and Communities	Ecology: Populations and Communities
Lesson 25—Ecosystems and the Biosphere	Ecology: Ecosystems and the Biosphere
Lesson 26—The Human Factor	Ecology: The Human Factor
	Behavior
	A Field Trip

LABORATORY KIT SUPPLIES

While many of the laboratory exercises provided for the telecourse *Cycles of Life: Exploring Biology* can be completed utilizing materials and equipment commonly available in the home, some of the exercises and experiments require specialized supplies. The following materials and equipment are included in the *Cycles of Life Laboratory Kit:*

Equipment:

Stainless steel forceps
Test tube clamp
Test tubes (4)
Glass slides (3)
Hand lens
Plastic squeeze bottle
Rubber stopper (2-hole)
Plastic tube
Rubber tube
Dialysis tubing

Chemicals:

Cobalt chloride paper
Silver nitrate
Distilled water
Benedict's solution
Tincture of iodine
Dextrose
Calcium oxide poser
Red litmus paper
Blue Litmus paper
PTC paper

Supplementary Color Plates

The *Cycles of Life Laboratory Kit* is available from Edutype: 8841 Crescent Drive, Huntington Beach, California 92646.

INTRODUCTION

The *Biology Laboratory Manual for the Telecourse Cycles of Life: Exploring Biology* provides distant students with structured laboratory experiences which can be completed at home. It gives detailed instructions for thirty-four laboratory exercises which are correlated to the telecourse lessons. For each exercise, the manual includes specific learning objectives, a list of the materials needed to complete the exercise, a discussion of the biological concepts involved, step-by-step instructions for conducting the exercises, and specific report sheets to be completed and submitted to the instructor.

In most of the exercises, the students utilize many materials commonly available in the home—salt, sugar, food coloring, fruits and vegetables—and common household implements and containers—pots, pans, water glasses, knives, newspaper, etc. To provide the specialized laboratory utensils and chemicals which are not available in the home, a simple laboratory kit has been developed which contains such items as test tubes, glass slides, plastic tubes and bottles and small amounts of basic chemicals such as *Benedict's Solution,* tincture of iodine, silver nitrate, etc. The kit also contains red and blue litmus paper and other indicators and *Supplementary Color Plates* selected for the course.

The manual also refers students to the telecourse textbook and to view visual images which would require a microscope, the *Photo Atlas for Biology,* published by Wadsworth Publishing Company, which contains more than six hundred full-color illustrations.

A MESSAGE TO THE INSTRUCTOR

Providing the distant learner with laboratory experiences to be completed outside of the traditional laboratory classroom is a challenge for the authors, for instructors who facilitate the learning experience, and for the student who must work independently. Every effort has been made to provide a well-balanced selection of exercises presented in a clear manner so that the distant learner will achieve the objectives with minimum student–instructor contact, although some contact is to be expected. Additionally, the number of exercises included will give the instructor choices so that exercises may be varied from term to term. With the variety of laboratory exercises available, the emphasis of the course could change for each term that the distance learning course is offered.

The last exercise, "A Field Trip," is included as a prototype so that the instructor may adapt the plan to a local museum, zoo, nature center or other facility that the instructor considers an important experience for students. Some of the most enthusiastic comments received from students by the authors have been related to similar field trip activities to natural areas and preserves.

The authors encourage suggestions and comments from instructors.

A MESSAGE TO THE STUDENT

Welcome to the laboratory component of the telecourse *Cycles of Life: Exploring Biology*. While the idea of completing a college-level laboratory science course may be new to you, it is not unique. An instructional television biology course was offered as early as 1980 by Tarrant County Junior College, Fort Worth, Texas, and it included a home-based laboratory component which is the forerunner of this manual. Several of the exercises from that original course, brought up-to-date, of course, are included in this manual, together with exercises developed specifically to fit the objectives of *Cycles of Life*.

It is extremely important that you follow all of the directions in each exercise so that you will achieve the appropriate results, and so that the materials in the laboratory kit will be sufficient to complete each exercise.

Follow all CAUTION messages and directions carefully to insure your safety!

Read all directions and discussion material BEFORE attempting to complete any exercise!

This manual has been written so that you will be able to achieve the state objectives for each laboratory experience with a minimum of additional assistance from your instructor; however, if you have any questions, please talk with your instructor before you proceed.

We believe that you will find this laboratory course experience very interesting and a valuable part of your science education.

1 THE SCIENTIFIC METHOD

LESSON OBJECTIVES

Upon completion of this laboratory exercise the student will be able to:

1. List six (6) steps which constitute the classical scientific method and briefly explain the significance of each as related to problem solving.

2. Demonstrate the ability to apply the scientific method by solving three (3) hypothetical problems.

MATERIALS NEEDED

1. Laboratory Manual
2. Pencil or pen
3. Thirty (30) leaves (see Exercise 1, Section D for collection method)
4. Transparent tape (Scotch tape)
5. Metric ruler

PREPARATION

Read the discussion which follows carefully before attempting to complete the exercise. Also read the appropriate chapter in your textbook.

DISCUSSION

Biology is the natural science which deals with the study of living things – plants and animals. Traditionally, biology was divided into two major areas: zoology, which dealt with the study of animals and botany which dealt with plants. Today biology is further divided in other more specialized fields of study including Anatomy, Biochemistry, Biophysics, Ecology, Genetics, Pathology, Physiology, Taxonomy, Virology, and others such as the new field of Exobiology, which proposes to study living things in space.

The vast amount of biological knowledge available today has been accumulated over the centuries by thousands of dedicated researchers. Research is usually conducted in an orderly, well–planned and carefully supervised manner. Two types of scientific research are recognized by scientists—basic or pure research and applied research. The basic or pure researcher is concerned with gaining new knowledge, for its own sake. If discoveries made through basic research do not offer immediate benefit to mankind, this does not concern the basic researcher. Applied research is concerned with finding ways of using scientific knowledge to develop useful products and technology to increase the welfare and comfort of mankind.

Regardless of the form of research, basic or applied, it involves an organized search for an answer to a question or problem. Research problems, scientists and available facilities vary so greatly that it would be a gross error to state that there is only one method by which scientific knowledge is advanced. However, it is helpful to consider the classical form of the scientific method since most investigations involve all or part of its procedures. In its purest form, the *Scientific Method* has six basic steps: observation, identification of the problem, collection of all known information regarding the subject, postulating a possible explanation, experimentation to test the postulate, and revision of the postulate as experimentation provides new facts.

Scientific investigations generally begin with the *observation* of some event, problem or phenomenon that stimulates a desire for knowledge and understanding. As the scientist makes observations, he/she must be careful to do so without prejudice or preconceived opinions which might result from previous experience or study. Ideally, the problem or phenomenon must be of such a nature that it can be observed and carefully measured as many times as necessary.

The second step involves asking questions such as "Under what conditions does the phenomenon occur?" or "What makes it happen?" To be experimentally valuable, the question must be one which can be tested. With these thoughts in mind, the researcher states as clearly as possible the *problem* he/she plans to investigate. Then a thorough inquiry is made to discover what is already known that is pertinent to the problem.

Next, the scientist tries, on the basis of available information, to explain the phenomenon or event by *postulating a hypothesis*. A hypothesis is a generalized statement which may or may not be the correct answer to the stated problem. The hypothesis must be submitted to testing and experimentation in an effort to verify or disprove it with information obtained from the experiments and tests.

Experimentation is designed to explore the validity of the hypothesis. The experiments must be carefully planned and conducted with precise instruments. Accurate records must be kept of every phase of the experiment. Most experiments consists of a *control group* and an *experimental group*. The experimental group is identical to the control group in every respect except one. The one substance, condition, etc that is being examined or tested is varied. All other factors are kept constant in both the control group and the experimental group. Therefore, the control serves as the basis or standard for evaluation of the result of changing the variable on the experimental results. Any experiment must be repeated many times under the exact same conditions to verify results. Other researchers must also be able to repeat the experiments under like conditions to insure validity of the results.

When all experiments are complete, the researcher must evaluate the results in an effort to reach a *conclusion*. The experimental results may support the original hypothesis or contradict it completely. (A negative result is often just as valuable to the researcher as an affirmative one.) In the latter case, the researcher can modify the original hypothesis, design other experiments and continue the research if such is indicated. On the other hand, the researcher may decide to abandon the project altogether.

In summary, the basic steps of the scientific method include:
1. Observation of a problem, event or phenomenon
2. Identification of the specific problem
3. Formulation of a hypothesis
4. Collection of available information
5. Design and execution of experimentation
6. Conclusion

LABORATORY EXERCISE PROCEDURE

Use the indicated steps of the Scientific Method with each of the following three (3) exercises. These exercises are designed to provide some experience using the scientific method as a tool to acquire knowledge. Carefully complete each exercise recording the required information on the report sheets. When complete, submit the reports sheets to the instructor for grading.

LABORATORY EXERCISES

I. You have observed that plants have leaves but that all plant leaves do not look alike. For example, leaves from oak trees are different from pecan tree leaves. You wonder if all the leaves on a single plant are alike and if each different plant has a unique leaf type. Since all the leaves on any one plant have a common origin, namely the same plant, you can hypothesize that all the leaves should be alike, but that each type of plant has its own unique type leaf.

   A. State the observation. _____
   _____
   _____

   B. What problem or question has been identified? _____
   _____
   _____

   C. The hypothesis is _____
   _____

   D. Experimentation: Select two different trees, shrubs or flowering plants. Collect 15 leaves from *each* plant (five leaves from the top of each plant, five leaves from the mid–section of each plant and 5 leaves from the bottom of each plant.) Carefully tape the leaves to the appropriate section of the report sheet, fill in the blanks in this section, and insert the sheets containing the leaves in a plastic bag. The bag will protect the leaves which must be submitted with the report sheet for grading. Carefully examine the 30 leaves. Record the observations in terms of width and length of each leaf; apparent thickness of each of the five leaves as compared to the other leaves from the same plant; color intensity of the leaves; and any other characteristic notes as the leaves are observed.

   E. Conclusion_____
   _____
   _____

   This exercise should be reported on Report Sheets 1–7.

The Scientific Method 3

II. You are employed in a large pharmaceutical laboratory and assigned to the task force searching for a cure for strain X pneumonia. In your haste to perform some tasks you accidently spill a vial of drug Y onto several cultures of strain X organisms. You are disgusted and tired and decide to go home and return early the next day to clean up the mess. In the morning, to your surprise, all of the strain X organisms in the cultures you spilled drug Y onto are dead.

Using the above information, complete Section II on the report sheet: A. State an observation, B. State the problem, C. Outline the hypothesis, D. Describe possible experiment(s) to the test the hypothesis, E. Evaluate the results of the hypothetical experiment(s), and F. State the conclusion(s). (Hint: Remember you should have a control group for comparison in your experiment.) This exercise should be reported on Report Sheets 7 and 8.

III. The Scientific Method is simply a logical method useful in solving any type of problem encountered in everyday situations. Think of some event, phenomenon, or everyday problem which is of interest to you. Some suggestions include purchasing a car, deciding on a career, choosing a college major or repairing an appliance. Show how you would apply the Scientific Method to the problem by stating your observation, problem, hypothesis, and explaining your possible experiment(s) in detail on the appropriate section of the report sheet. Does your data support your hypothesis or disprove it? Why? How? This exercise should be reported on Report Sheet 9.

Complete the report sheet and submit it to the instructor for grading.

THE SCIENTIFIC METHOD

Report Sheet 1

Name _____

Student ID # _____

Campus _____

Date _____

I.   A.   Observation _____

_____

_____

     B.   Problem _____

_____

_____

     C.   Hypothesis _____

_____

_____

     D.   Collected information _____

_____

_____

     E.   Experiment _____

_____

_____

     F.   Conclusion _____

_____

_____

The Scientific Method　　　　　　　Student ID # _____
Report Sheet 2

1. Attach 15 leaves from the same plant to this sheet recording the appropriate information as indicated. Plant A leaves are from a _____(what kind of tree, shrub, plant?)

    a. Attach the 5 leaves from the top of the plant here. Measure the length and width of each leaf in centimeters.

Leaf	1	2	3	4	5
Length	___	___	___	___	___
Width	___	___	___	___	___

*Average length of top leaves _____ cm

*Average width of top leaves _____ cm

    b. Attach the 5 leaves from the mid–section of the plant here. Measure the length and width of each leaf in centimeters.

Leaf	1	2	3	4	5
Length	___	___	___	___	___
Width	___	___	___	___	___

*Average length of mid–section leaves _____ cm

*Average width of mid–section leaves _____ cm

The Scientific Method
Report Sheet 3

Student ID # _____

    c.   Attach the 5 leaves from the bottom of the plant here. Measure the length and width of each leaf in centimeters.

Leaf	1	2	3	4	5
Length	_____	_____	_____	_____	_____
Width	_____	_____	_____	_____	_____

*Average length of bottom leaves _____cm

*Average width of bottom leaves _____cm

*Average length or width of leaves is obtained by adding the 5 lengths or widths together and dividing the total obtained by 5

    d.   Complete the following chart after careful observation of the leaves

	Top Leaves	Mid–Section Leaves	Bottom Leaves
Average Length			
Average Width			
Leaf Color **			
Leaf Thickness **			
Other			

    ** Use 1 to indicate darkest color or thickest leaf
        Use 2 to indicate an intermediate color or thickness
        Use 3 to indicate lightest color or thinnest leaf

The Scientific Method Student ID # _____
Report Sheet 4

2. Attach the 15 leaves from the other plant to this sheet recording the appropriate information as indicated. Plant B leaves are from a _____(what kind of tree, shrub, plant?)

   a. Attach the 5 leaves from the top of the plant here. Measure the length and width of each leaf in centimeters.

Leaf	1	2	3	4	5
Length	_____	_____	_____	_____	_____
Width	_____	_____	_____	_____	_____

*Average length of top leaves _____cm

*Average width of top leaves _____cm

   b. Attach the 5 leaves from the mid–section of the plant here. Measure the length and width of each leaf in centimeters.

Leaf	1	2	3	4	5
Length	_____	_____	_____	_____	_____
Width	_____	_____	_____	_____	_____

*Average length of mid–section leaves _____cm

*Average width of mid–section leaves _____cm

8   Exercise 1

The Scientific Method Student ID # _____
Report Sheet 5

    c. Attach the 5 leaves from the bottom of the plant here. Measure the length and width of each leaf in centimeters.

Leaf	1	2	3	4	5
Length	_____	_____	_____	_____	_____
Width	_____	_____	_____	_____	_____

*Average length of bottom leaves _____ cm

*Average width of bottom leaves _____ cm

*Average length or width of leaves is obtained by adding the 5 lengths or widths together and dividing the total obtained by 5

    d. Complete the following chart after careful observation of the leaves

	Top Leaves	Mid–Section Leaves	Bottom Leaves
Average Length			
Average Width			
Leaf Color **			
Leaf Thickness **			
Other			

    ** Use 1 to indicate darkest color or thickest leaf
        Use 2 to indicate an intermediate color or thickness
        Use 3 to indicate lightest color or thinnest leaf

The Scientific Method Student ID # _____
Report Sheet 6

3. Complete the following charts:

Leaf Length*

	Top Leaves	Mid–Section Leaves	Bottom Leaves
Plant A	_____	_____	_____
Plant B	_____	_____	_____

Leaf Width*

	Top Leaves	Mid–Section Leaves	Bottom Leaves
Plant A	_____	_____	_____
Plant B	_____	_____	_____

Leaf Color*

	Top Leaves	Mid–Section Leaves	Bottom Leaves
Plant A	_____	_____	_____
Plant B	_____	_____	_____

Leaf Thickness*

	Top Leaves	Mid–Section Leaves	Bottom Leaves
Plant A	_____	_____	_____
Plant B	_____	_____	_____

The Scientific Method Student ID # _____
Report Sheet 7

4. Answer the following questions using the observations.

    a. Are the leaves from Plant A and Plant B alike? _____ How do they differ?

       _____
       _____
       _____

    b. Are all of the leaves from Plant A alike? _____ How do they differ?

       _____
       _____
       _____

    c. Are all of the leaves from Plant B alike? _____ How do they differ?

       _____
       _____
       _____

    d. Could this experiment be repeated exactly by another researcher? _____
       Why is the repeatability of experimentation important?

       _____
       _____
       _____

II. A. Observation _____

       _____
       _____
       _____

The Scientific Method
Report Sheet 8

Student ID # _____

B. Problem _____
   _____
   _____
   _____

C. Hypothesis _____
   _____
   _____
   _____

D. Possible Experiment(s) _____
   _____
   _____
   _____

E. Results of Experiment(s) _____
   _____
   _____
   _____

F. Conclusion _____
   _____
   _____
   _____

G. Did the information gained from the experiment support the hypothesis?_____
   How? _____

The Scientific Method Student ID # _____
Report Sheet 9

III. A. Observation _____-_____

_____

_____

_____

B. Problem _____

_____

_____

_____

C. Hypothesis _____

_____

_____

_____

D. Possible Experiment(s) _____

_____

_____

_____

E. Results of Experiment(s) _____

_____

_____

_____

F. Conclusion _____

_____

_____

_____

The Scientific Method Student ID # _____
Report Sheet 10

G. Did the information gained from the experiment support the hypothesis _____

How? _____

_____

_____

2   BASIC CHEMISTRY

LESSON OBJECTIVES

Upon completion of this laboratory exercise the student will be able to:

1. Define matter and energy.

2. Determine the atomic number and atomic mass number of an element.

3. Using the atomic number and atomic mass number, determine the number of protons, neutrons, and electrons in an atom of an element.

4. Explain the difference between a physical change and a chemical change and give examples of each.

5. Classify reactions as:
   a. synthesis (condensation)
   b. decomposition (hydrolysis)

6. Determine whether certain solutions are acidic or basic.

7. Complete the written parts of this exercise.

MATERIALS NEEDED

1. Laboratory Manual
2. Wooden matches (kitchen matches)
3. Test tubes
4. Test tube holder
5. Red litmus paper*
6. Blue litmus paper*
7. Glass slide*
8. Table salt (sodium chloride—NaCl)
9. White vinegar (acetic acid)

* These items are found in your lab kit.

PREPARATION

Read the discussion which follows carefully before attempting to complete the exercise. Also read the appropriate chapter in your textbook.

DISCUSSION

All living things consist of *matter*, which is anything that occupies space and has mass. Matter exists in three states: solid, liquid, and gas. Matter can be changed from one state to the next by the addition or removal of energy. *Energy* is the capacity to do work, to put mass into motion. Mass and energy can be neither created nor destroyed, however, one can be converted into the other. All forms of matter are made up of a limited number of building blocks called chemical elements. *Chemical elements* are substances that cannot be split into simpler substances by ordinary chemical means. Over a hundred chemical elements are currently recognized with 92 of them occurring naturally. (The remaining ones can be artificially created.) The elements are given letter abbreviations and are arranged in a table, called the periodic table, based on their chemical natures.

Each element is made up of atoms. *Atoms* are the smallest units of matter that enter into chemical reactions. Atoms are composed of even smaller units called protons, electrons, and neutrons. The protons and neutrons are located in a central unit called the nucleus. Each proton and neutron has an atomic mass of one unit. Each proton carries a positive electrical charge; each neutron is neutral. The electrons have a mass equal to 1/2000 the mass of either one proton or one neutron. As a result, the mass of the electrons are not considered when determining the mass of an atom of an element. The electron does carry a negative electrical charge.

The number of electrons in an atom of an element always equals the number of protons. Since each electron carries a single negative charge and each proton carries a single positive charge, the electrons and protons of an element balance the electrical charges resulting in an atom that is electrically neutral. This holds true for atoms that are not undergoing a chemical reaction; that is, atoms that are not reacting with other atoms. The number of protons in an atom is called the *atomic number*.

The *mass number* of an atom is the total number of protons and neutrons in the atom. Atoms of an element, however, may have different mass numbers because they have different numbers of neutrons. Atoms of an element that differ in the number of neutrons from the majority of the atoms of an element are called isotopes. All isotopes of an element have the same chemical properties because they have the same number of electrons, since the chemical properties of an atom are a function of its electrons.

When atoms combine with or separate from other atoms, a chemical reaction is said to have occurred. A chemical reaction results in new products with different chemical properties being formed. All life processes involve chemical reactions, and electron interactions are the basis of all chemical reactions.

Electrons surround the nucleus of an atom in various energy levels, ranging from the first energy level immediately surrounding the nucleus to more far-ranging levels. The number of electrons that can occupy any given energy level is based upon the shape of that level and the total number of electrons found in that atom. Each energy level has a maximum number of electrons that it can hold. For an atom to be stable, the outermost energy level must be filled with its maximum number of electrons. Atoms tend to either empty their outermost energy level and fall back to the next filled level or fill the outermost level to its maximum. An atom that contains half of the necessary number of electrons in the outermost level necessary to fill that level may share its electrons with other atoms to fill the outermost energy level. The loss and gain of electrons and the sharing of electrons result in the formation of chemical bonds. Atoms are combined by chemical bonds to form molecules. A *molecule* contains two atoms of the same kind or two or more different kinds of atoms. A *compound* is a substance that can be broken down into two or more different elements by chemical means. The molecules of a compound always contain atoms of two or more different elements.

Chemical reactions are simply the making or breaking of bonds between atoms. After a chemical reaction, the total number of atoms is the same, but because they are rearranged, there are new molecules with new properties. Most chemical reactions can be classified into four major types. These are (a) *synthesis* or condensation, (b) *decomposition* or hydrolysis, (c) *replacement*, either single or double, where one or more atoms exchange places with other atoms in the reaction, and (d) *equilibrium reactions* in which the reaction is reversible so that reactants can be recovered.

Atoms are electrically neutral because the number of positively charged protons equals the number of negatively charged electrons. When a atom gains or loses electrons trying to achieve stability with a filled outermost energy level, the electrical balance is upset. If the atom gains electrons, it acquires a negative charge; if it loses electrons, it has a positive charge. Such negatively or positively charged particles are called *ions*. Ions in solution (a water solution) are called electrolytes because the ionic solution is capable of conducting an electric current.

One of the most significant ions is the hydrogen ion because of its reactivity. Compounds that separate into one or more hydrogen ions and one or more anions (negative ions) are called *acids*. Compounds that separate into more hydroxide ions and one or more cations (positive ions) are called *bases*. A solution's acidity or alkalinity is expressed on the pH scale which runs from 0 to 14. This scale is based on the concentration of hydrogen ions in a solution. To control the concentration of hydrogen ions, buffer systems are used. The function of a buffer system is to convert strong acids or bases (relatively unstable, ionize easily) into weak acids or bases (relatively stable, do not ionize easily). This insures that strong acids or bases do not alter pH very much. The chemicals that replace strong acids or bases with weaker ones are called *buffers*.

LABORATORY EXERCISE PROCEDURE

Carefully follow the directions for each of the following experiments, recording observations, conclusions and answers on the appropriate section of the report sheet. Read each laboratory exercise and assemble all needed materials before attempting the experiments. Observe all cautions. When the experiments are complete submit the completed report sheet to the instructor for grading.

LABORATORY EXERCISES

I. Atomic number is the number of protons in an atom of an element. Atomic mass number is the number of protons plus neutrons in an atom of an element. In a chemically inert atom, the number of protons equals the number of electrons. Using the textbook as a reference and the periodic table determine the atomic number, atomic mass number, number of protons, and electrons for the following elements: oxygen, carbon, hydrogen, phosphorus and nitrogen. Record this information on Report Sheet 1.

II. Physical change is the transformation of a substance in which only a physical property such as color, odor or phase (solid, liquid, gas) changes. A chemical change results in the production of new substances which may or may not have the same physical and chemical properties of the original substance.

Classify each of the following reactions as a physical or chemical change. Record your results on the Report Sheet.

    A. Obtain a wooden match. Strike the match and let it burn for a few moments. **CAUTION: BURNING MATCHES ARE HOT AND CAN CAUSE FIRES.**

    B. Mix some table salt in tap water. Place 2–3 drops on a glass slide. Allow the slide to ir dry. After the liquid has evaporated, taste the **residue. CAUTION: NEVER TASTE A CHEMICAL UNLESS SPECIFICALLY TOLD TO DO SO.**

III. Litmus paper can be used to determine whether a solution is acidic or basic. An acid turns blue litmus red and does not change red litmus. A base turns red litmus blue and does not change blue litmus. Determine whether each of the following solutions is acidic or basic by putting a drop of each solution on , a strip of red litmus paper and a strip of blue litmus paper and observing any color change. **CAUTION: VINEGAR AND BLEACH CAN BE HARMFUL TO THE SKIN. TRY TO AVOID SKIN CONTACT. DO NOT LET VINEGAR AND BLEACH COME IN CONTRACT WITH EACH OTHER.** You may try other solutions such as orange juice, tomato juice, or other solutions that are available. Record your results on Report Sheet 2.

BASIC CHEMSITRY

Report Sheet 1

Name _____

Student ID # _____

Campus _____

Date _____

I. Protons, Neutrons, and Electrons

Element	Atomic Number	Mass Number	Protons	Electrons	Neutrons
oxygen (O)					
carbon (C)					
hydrogen (H)					
phosphorus (P)					
nitrogen (N)					

II. Physical versus Chemical Changes

    A. Describe the changes the match underwent. _____

_____

        Is this a physical or a chemical change? _____

        Why did you classify it as you did? _____

_____

_____

    B. Describe what happened when the salt and water were mixed. _____

_____

        Describe the residue after the water evaporated. _____

_____

The Scientific Method　　　　　　Student ID # _____
Report Sheet 2

What did it taste like? _____

Does this residue represent a physical or a chemical change? _____

Why did you classify it as you did? _____

_____

III. Acid–Bases

　　A.　pH of Common Substances

Substance	Color of Blue Litmus	Color of Red Litmus	Is this substance acidic or basic?
saliva			
urine			
vinegar			
bleach			
water			

3   THE CELL AND CELL PROCESSES

LESSON OBJECTIVES

This exercise examines the structure of the cell and the functions of the organelles of the cell. At the completion of this laboratory, the student should be able to:

1. State the cell theory.

2. Define ultrastructure and cell organelle.

3. Label a drawing of a typical cell using the terms cell membrane, nucleus, nucleolus, nuclear membrane, endoplasmic reticulum, ribosome, mitochondria, lysosome, Golgi apparatus, cell wall, plastid, vacuole and/or centrioles.

4. Explain the function of each of the terms listed in objective 3.

5. Explain the difference between a typical animal cell and a typical plant cell.

6. List the factors that affect the rate of diffusion.

7. Explain diffusion and osmosis.

MATERIALS NEEDED

1. 3 transparent 8-ounce containers such as drinking glasses. (NOTE: One must be heat resistant.)
2. 6 ounces of ice water
3. 6 ounces of very hot water
4. Food coloring
5. Clock or watch with second hand
6. Cellophane tubing (obtained at Orientation or in the lab kit)
7. Sugar
8. Salt
9. Drinking straw (very slender like a small cocktail straw)
10. Pencil
11. Twine
12. Small pan such as a pie pan
13. Red beet, white potato, carrot (If a fresh red beet is not available, a canned red beet may be substituted.)

PREPARATION

Read the discussion which follows carefully before attempting to complete the exercise.
Also read the appropriate chapter in your textbook.

DISCUSSION

Cell Structure

The cell is the structural and functional unit of life. Every living thing is made up of cells. The cytologist (one who studies cells) is limited by the light microscope to seeing cell organelles such as the nucleus, cell membranes, vacuoles, and chloroplasts. The electron microscope permits investigation of the cell ultrastructure, the fine structure of the cell which cannot be seen with the light microscope.

The term "cell" was coined by Robert Hooke, who first identified the structure by looking at cork cells under a microscope. Robert Brown is credited with naming the cell nucleus. The cell theory is credited to Schleiden and Schwann. The cell theory states that all living things are made up of cells or materials made by cells and that all cells come from preexisting cells. Most of the current information about cell structure has come from technological advances in instrumentation. The electron microscope and ultracentrifuge are two of the more recently developed instruments that have aided scientists in their study of the cell.

There are many different kinds of cells; however, all cells are very similar in most important respects. Animal cells generally vary in shape, taking their shape from their surroundings. On the other hand, plant cells have a definite shape determined by the cell wall which surrounds them. Most cells are limited in size, many between 10–30 micrometers in diameter. (A micrometer = 1/25,000 inch.) Cells that show a very low level of cell organization with nuclear material not enclosed in a membrane are called **procaryotic**. These include bacteria and blue-green algae. Cells that are highly organized with a well-defined nucleus bounded by a nuclear membrane are termed **eukaryotic** cells. Eukaryotic cells contain various organelles which have specific functions to perform for the cell.

The basic unit of cell structure is the unit membrane. The membrane surrounds the cell; divides it into compartments; and marks the outer limits of individual cellular components such as the nucleus, mitochondria, lysosomes, cavities of the endoplasmic reticulum and the Golgi apparatus. The function of the membrane is to provide selective barriers controlling the amount and nature of substances passing between the cell and its environment and between intracellular components. The membrane is composed of a lipid matrix interspersed with protein – somewhat like a chocolate chip cookie with the protein being the chips.

The **cell membrane** is present at the outer surface of all cells. It is the primary barrier that determines what can enter and leave the cell. It is a selectively permeable membrane (one which allows free passage of some substances while other substances cannot cross the membrane). All the material between the cell membrane and the nuclear membrane is termed **cytoplasm**. Cytoplasm contains all the cell organelles and surrounds the **nucleus**.

The nucleus is the information and control center for the cell. It is bounded by a unit membrane called the **nuclear membrane** and contains the genetic instructions which control and regulate cell division and all cellular processes. It also contains one or more **nucleoli**. A nucleolus is the site of ribosome formation in the nucleus.

The primary organelles within the cytoplasm are the endoplasmic reticulum, Golgi apparatus, mitochondria, lysosomes, and ribosomes. The **endoplasmic reticulum** (ER) is a highly developed and extensive series of membranes that traverse the cytoplasm. The ER connects various parts of the cell and contains enzymes which play important roles in metabolic sequences. **"Rough" ER** contains **ribosomes**. Ribosomes are also found free in the cytoplasm. The function of ribosomes is protein synthesis. Ribosomes are particularly abundant in cells synthesizing large quantities of protein to be used by the cell for growth and other internal functions. The **Golgi apparatus** is the packaging system in the cell. Materials secreted by the cell are concentrated and packaged by the Golgi apparatus into vesicles and vacuoles. The vesicles and vacuoles are later released at the cell surface or used in the cell itself. The **mitochondria** produce energy for the cell. Cell respiration and the production of chemical energy in the form of ATP occurs in the mitochondria. The "digestive" system of the cell is the **lysosome**. The lysosome is a membrane sac containing lytic enzymes capable of hydrolyzing (breaking down) all classes of macromolecules in cells. It serves to rid the cell of extra material or potentially toxic materials taken into the cell.

Specialized cell structures include the cell wall, plastids, and vacuoles. The **cell wall** is found only in plant cells and bacteria. The function of the cell wall is support and protection. The cell wall is located outside the cell membrane and gives the plant cell its shape. The cell wall is composed mostly of cellulose and is non–living. **Plastid**s are membrane–lined sacs which usually contain pigments. The most specialized plastid is the **chloroplast** which is the site for photosynthesis in the plant cell. Other plastids include chromoplasts which may be converted into vitamins when taken into the body. Sweet potatoes, squash and carrots contain chromoplasts which are sources of carotenes that may be converted into vitamin A in the body. Such carotenes are termed pro–vitamin A. Plastids are found in plant cells only. **Vacuoles** are spaces in cytoplasm which store food reserves or liquid. In the plant a large central vacuole aids in plant turgor (the basis of mechanical support in most flowers, leaves and succulent young stems related to the movement of water into the cells of the plant). A large central vacuole is characteristic only of plant cells.

Cell Processes

Movement of material into and out of the cell is controlled by the cell membrane. The cell membrane is selectively permeable; that is, it permits some molecules to pass through and inhibits others. Two types of movement are associated with the cell membrane—**passive transport** and **active transport**.

Passive transport of molecules does not require an expenditure of energy. The most important types of passive transport are diffusion and osmosis. **Diffusion** is the movement of molecules of a particular kind from an area of their higher concentration to areas of their lesser concentration. **Osmosis** is the diffusion of water molecules from an area of their higher concentration through a selectively permeable membrane to an area of lower water concentration.

Active transport requires an expenditure of energy in order to move materials across the membrane. In active transport materials may move from an area of their *lower* concentration to an area of higher concentration.

The cell membrane is a *selectively* permeable membrane; thus, a cell is affected by the concentration of solutions. A cell may find itself in an isotonic, hypotonic, or hypertonic solution. An **isotonic solution** is one in which the water concentration is equal on both sides of the cell membrane. A **hypotonic solution** is one in which the concentration of water molecules is higher on the outside of the cell than inside the cell. Water will flow into the cell to equalize the concentrations of water molecules, causing the cell to swell and possibly burst. A **hypertonic solution** exists when the concentration of the water molecules outside the cell is lower than the concentration of water molecules inside the cell. Water will leave the cell to equalize the concentration of water molecules outside the cell. This causes the cell protoplasm to decrease in volume. The resulting shrinking or collapse of the cell is term plasmolysis.

LABORATORY EXERCISE PROCEDURE

Carefully follow the directions for each of the following experiments, recording observations, conclusions and answers on the appropriate section of the report sheet. Read each laboratory exercise and assemble all needed materials before attempting the experiments When the experiments are complete submit the completed report sheet to the instructor for grading. Also read the appropriate chapter in your textbook.

LABORATORY EXERCISES

I. On the report sheet is a drawing of a typical animal cell. Using your textbook and this exercise as references, answer the appropriate questions pertaining to the cell, its structures and their functions.

II. The purpose of this exercise is to demonstrate that the rate of diffusion is greatly affected by temperature. Materials needed include 2 eight ounce containers, ice water, boiling water, food coloring, and a watch or clock with a second hand.

Prepare two containers of water, one with six ounces of very hot water and one with six ounces of ice water. Add three drops of food coloring to each glass. Compare the rate of diffusion in the two containers. Measure the length of time for complete diffusion of the food coloring throughout the hot water and for complete diffusion throughout the cold water. Record your measurements in minutes and seconds on the report sheet Exercise II.

III. The purpose of this exercise is to investigate osmosis as it occurs across a semi-permeable membrane represented by cellophane tubing. Materials needed for this exercise include cellophane tubing, tap water, sugar, food coloring, some twine string, a drinking straw, a pencil, a glass container, a ruler and a small cup.

Mix 4 tablespoons of tap water, 1 tablespoon of sugar and 2–3 drops of food coloring in a cup and set aside. Soak the cellophane tubing in a bowl of water for 2–3 minutes before proceeding. Then gently rub the tubing between two fingers until it "opens." Securely tie one end of the cellophane tubing with the twine string. After tying one end of the cellophane tubing, pour some of the colored sugar water into the cellophane tubing bag. Insert a drinking straw through the open end of the bag, making sure it extends well into the liquid. Tie the open end securely but be careful not to "pinch in" the straw. Using more twine and a pencil, suspend the bag and straw in a glass of water. Make sure that the water in the bag is level with the water in the glass. (See Figure 1.)

Using the ruler, measure the distance from the table top to the water level in the glass and in the straw. Repeat this measurement at the prescribed intervals on the report sheet. Answer all questions on the report sheet relating to Exercise III before proceeding.

IV. The membranes of both plant and animal cells are selectively permeable; that is, some substances cannot pass through them, others can pass through slowly, and still other substances can pass freely in and out of the cell. Selective membrane permeability can be demonstrated with a red beet. The red color of the beet is due to the presence of red anthocyanin pigment contained in the cells. Materials needed include a red beet, quart jar, tap water, knife to slice the beet, and a small cooking pan.

Slice the beet into 1/8-inch thick slices. Place the slices in a jar. Pour 1-1/2 cups of tap water at room temperature over the beets. Allow the beet slices to remain in the water overnight. Observe the beet slices and the water. Answer the appropriate questions on the report sheet before proceeding. A canned beet can be substituted for the fresh beet although the results will not be as dramatic. (If using canned beets, rinse thoroughly before putting in the tap water overnight.)

Next drain the beet slices and place them in 1-1/2 cups of fresh tap water. Bring the water to a boil and boil for 15 minutes. The heat will destroy the cell membranes. Carefully remove the beet slices from the boiling water and observe. Answer the appropriate questions on the report sheet.

V. Osmosis in plant tissue can be observed by a simple experiment. Materials needed include a medium size white potato, small knife or cork borer, pan or bowl, tap water and sugar.

The potato is a tuber which is a modified stem. Cut both ends from a medium sized potato so that the cut surfaces are parallel to each other. Using a small knife or cork borer, make a pit in one of these surfaces about three-fourths of the length of the potato. The cavity or pit now forms a "cup" which you should half fill with dry sugar. Set the potato in a pan or bowl of water but first peel the portion that is submerged in water. After allowing the demonstration to set for 3 to 4 hours, carefully observe any changes that have occurred. Record your observations on the report sheet, answering the appropriate questions.

VI. Repeat the above demonstration substituting a carrot, which is a tap root, for the potato. A small bottle, jar or drinking glass should be substituted for the pan of water. After 3–4 hours, observe the sugar in the pit. Answer the appropriate questions on the report sheet.

VII. Plasmolysis of plant tissue can be demonstrated easily. Materials needed include 2 thin slices of white potato, 2 bowls, water and salt. Place one slice of potato in a bowl containing one cup of tap water. Place the other slice of potato in a bowl containing one cup of tap water plus salt. (Add salt to the cup of water until no more salt will dissolve in the water. This is a saturated salt solution.) Allow both potato slices to remain in their respective bowls for one hour. Remove the potato slices and compare their texture. Answer the appropriate questions on the report sheet.

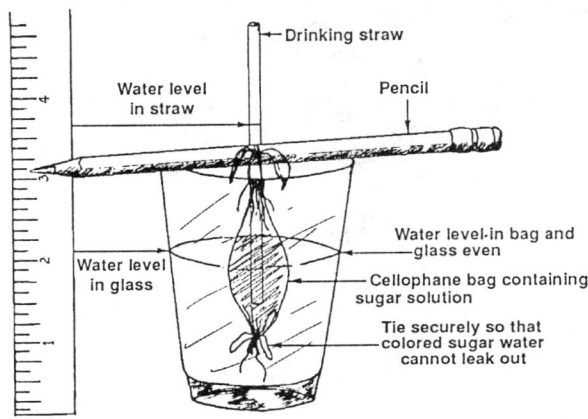

FIGURE 1

THE CELL AND CELL PROCESSES

Report Sheet 1

Name _____

Student ID # _____

Campus _____

Date _____

I. A. Label the drawing of the typical animal cell using the following labels: cell membrane, nucleus, nucleolus, endoplasmic reticulum, centrioles, mitochondrian, vacuole, ribosomes.

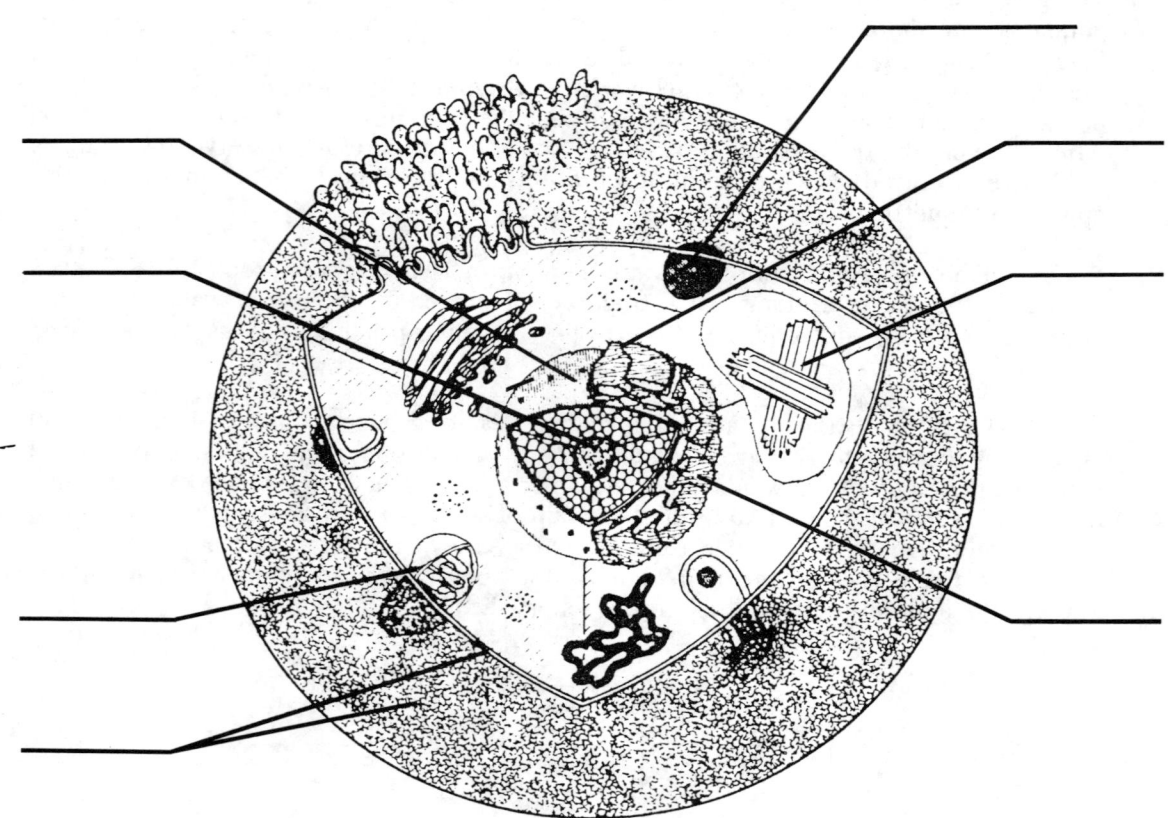

FIGURE 2

26   Exercise 3

The Cell and Cell Processes   Student ID # _____
Report Sheet 2

B. State the cell theory _____

_____

C. Define ultrastructure_____

_____

D. Define cell organelle _____

_____

E. Give the function for each of the following:

   Cell Structure   Function

   1. Cell membrane   _____
   2. Nucleus   _____
   3. Nucleolus   _____
   4. Nuclear membrane   _____
   5. ER   _____
   6. Ribosome   _____
   7. Mitochondrion   _____
   8. Lysosome   _____
   9. Golgi apparatus   _____
   10. Cell wall   _____
   11. Plastid   _____
   12. Vacuole   _____

The Cell and Cell Processes Student ID # _____
Report Sheet 3

    F. Compare a typical plant cell and a typical animal cell. (Refer to the textbook for further information.)

       Plant Cell   Animal Cell

      1. _____

      2. _____

      3. _____

      4. _____

II. 1. Diffusion occurs more quickly in _____ (hot, cold) water.

    2. Length of time for diffusion to occur in hot water = _____

    3. Length of time for diffusion to occur in cold water = _____

III. 1.
                          Water level in straw   Water level in glass
                        measured from table top   measured from table top

Start of experiment_____

10 minutes_____

15 minutes_____

20 minutes_____

    2. What changes in the water level in the glass did you observe? _____

_____

In the water level in the straw?_____

_____

Explain why these changes occurred._____

_____

_____

28   Exercise 3

The Cell and Cell Processes   Student ID # _____
Report Sheet 4

        Did the water move more rapidly into the bag and up into the straw or out of the bag

        _____

        How does the cellophane function in the exercise?_____

        _____

IV. Describe the appearance of the beet slices left in water at room temperature overnight.

        _____

        _____

        Of the beet slices boiled for 15 minutes._____

        _____

        Describe the water the beet slices were in after the experiment:

        The room temperature water_____

        _____

        The boiling water_____

        _____

        Did the boiling water render the cells more or less permeable? _____

        What experimental evidence supports this conclusion?_____

        _____

V.   What change occurred in the sugar in the cavity of the potato?_____

        _____

VI. After several hours, was the sugar in the carrot dry?_____

        What conclusion(s) regarding the movement of water through a root or stem can be made as a result of these two demonstrations?

        _____

The Cell and Cell Processes Student ID # _____
Report Sheet 5

VII. Describe the difference observed in the texture of the two potato slices.

_____

_____

Explain what happened to the potato slice in the salt solution to cause the observed change.

_____

_____

Is the salt solution more nearly isotonic, hypotonic, or hypertonic with respect to the cells of the potato?

_____

Is the water more nearly isotonic, hypotonic, or hypertonic with respect to the cells of the potato?

_____

In this experiment, what was the control? _____

_____

4 PHOTOSYNTHESIS

LESSON OBJECTIVES

Upon completion of this laboratory exercise the student will be able to:

1. Define photosynthesis and explain its importance.

2. Describe the structure and function of a chloroplast.

3. Describe the reactions of photosynthesis and describe where these reactions occur in he chloroplast.

4. List the raw materials and products and describe the processes involved in the two major reactions of photosynthesis.

5. List the uses of the products of photosynthesis.

6. Contrast photosynthesis to cellular respiration.

MATERIALS NEEDED

1. Laboratory Manual
2. Pencil or pen
3. *Photo Atlas for Biology*
4. Two small potted plants

PREPARATION

Read the discussion which follows carefully before attempting to complete the exercise. Also read the appropriate chapter in your textbook.

DISCUSSION

All living organisms require energy to conduct the many functions necessary to sustain life. The source of this energy for cells is the chemical bonds of food molecules. Some organisms such as green plants and algae are capable of trapping sunlight energy and converting it to chemical bonds in food molecules. Other organisms such as animals depend upon already produced food molecules for their energy. Organisms that are able to make their food molecules from sunlight and inorganic raw materials are **autotrophs** (self–feeders). Organisms that eat food to obtain energy are termed **heterotrophs** (different feeders).

The process of converting sunlight energy to chemical–bond energy is called **photosynthesis**. **Cellular respiration** is the process whereby during a series of reactions cells release the chemical–bond energy from food and convert it into usable forms. All organisms must be able to make this energy conversion regardless of the source of their food molecules. Whether organisms manufacture food or take it from the environment, they use food molecules as a source of energy. This process of energy conversion is essentially the same in all organisms (see Figure 1).

Within cells, specific biochemical pathways are carried out in specific organelles. These organelles contain specific enzymes necessary to control the reactions and the surfaces upon which the chemical reactions occur. **Chloroplasts** are the site of photosynthesis, and **mitochondria** are the site of most of the reactions of cellular respiration. There are some exceptions, however. Prokaryotic cells lack mitochondria and chloroplasts, yet some are capable of processes very similar to the cellular respiration and photosynthesis process found in eukaryotic cells. This exercise will concentrate on chloroplasts and their role in photosynthesis.

For most plants, the entire process of photosynthesis takes place in the leaf, where the cells contain large numbers of organelles called chloroplasts. Chloroplasts are oblong membrane sacs containing many thin, flat disks called **thylakoids**. The thylakoids contain the chlorophyll pigments, other pigments, electron transport molecules, and enzymes. The thylakoids are stacked in groups called **grana** (singular granum). The regions between the grana are called the **stroma** of the chloroplast. Photosynthesis is divided into two stages with the first stage, the **light–energy conversion stage,** taking place in the grana. During this stage, the light energy from the sun is converted into chemical–bond energy. The second stage of photosynthesis, the **carbon dioxide conversion stage**, occurs outside the grana in the stroma. During this stage, carbon dioxide becomes incorporated into a simple sugar molecule. The simple sugar molecules that are produced form a complex sugar molecule called starch. The starch is stored in the plant for future energy needs, is broken down to simple sugars for use by the plant, and/or is consumed by animals for their energy needs.

In summary, photosynthesis involves plants capturing the energy of the sun and converting it to stable chemical energy. Overall, the reactions of photosynthesis convert 6 molecules of carbon dioxide and 6 molecules of water to 1 molecule of glucose (sugar) and 6 molecules of oxygen. Glucose is an excellent energy storage molecule found in or used by all organisms. Energy is released as ATP (adenosine triphosphate) when glucose is broken down to carbon dioxide and water. Cellular respiration reactions retrieve energy stored in the bonds of the products of photosynthesis. In the process of these chemical reactions, plants provide oxygen to the atmosphere as they convert sunlight and carbon dioxide into energy molecules. Animals use the energy molecules (simple sugars) to supply their energy needs and in the process supply the atmosphere with carbon dioxide. The energy cycle is thus maintained.

LABORATORY EXERCISE PROCEDURE

Carefully follow the directions for each of the following experiments, recording observations, conclusions and answers on the appropriate section of the report sheet. When the exercise is complete, submit the completed report sheet to the instructor for grading. Also read the appropriate chapter(s) in your textbook.

LABORATORY EXERCISES

I. Define the indicated terms on the lab report sheet.

II. The Leaf

    A. Draw the structure of the green leaf, label the parts.

    B. Draw and label the parts of the vascular bundles.

    C. Answer all questions related to the leaf.

III. Take the two potted plants and place one in a closet or other dark place and one in bright sunlight. Leave them for at least 4 days. On the fourth day (or later), observe the appearance of the two plants. Answer the questions in the lab report.

IV. Photosynthesis and the Chloroplast

    A. Identify the plant structures that are involved in photosynthesis. Draw and label the parts of the chloroplast.

    B. Identify the two phases of photosynthesis, where the phases occur, and the end products of the two phases.

    C. Answer all questions in the lab report.

PHOTOSYNTHESIS

Report Sheet 1

Name _____

Student ID # _____

Campus _____

Date _____

I. Define the following terms:

   A. Photosynthesis _____
   _____
   _____

   B. Cellular respiration _____
   _____

   C. Chloroplast _____
   _____

   D. Mitochondrion _____
   _____

   E. Autotroph _____
   _____

   F. Heterotroph _____
   _____

II. A. Draw the structure of the leaf (Use *Photo Atlas* pp. 81–82 as a guide). Label the following structures: upper epidermis, palisade layer, chloroplasts, spongy layer, stomates, and epidermal hairs.

34 Exercise 4

Photosynthesis Student ID # _____
Report Sheet 2

    B.   Draw and label the parts of the vascular bundle. (Refer to fig. 82b, pg. 82, of the *Photo Atlas*.)

    C.   Answer the following questions:

        1.   Identify the function(s) of the following:

            a.   Upper epidermis _____

            b.   Chloroplasts _____

            c.   Spongy layer _____

            d.   Stomata _____

        2.   What is the role of the guard cells of the stomata? _____

        3.   What is the role of epidermal hairs? _____

Photosynthesis Student ID # _____
Report Sheet 3

    4. What is the role of phloem? _____

_____

       Of xylem? _____

_____

III. Describe the appearance of the potted plant placed in a dark area. _____

_____

_____

Describe the appearance of the potted plant placed in sunlight. _____

_____

_____

What does the experiment with light prove? _____

_____

_____

IV Draw the structure of the chloroplast (use the *Photo Atlas*, pg. 8, fig. 8e as a guide) Label the following structures: grana, intergrana, stroma, and thylakoid disk.

36 Exercise 4

Photosynthesis Student ID # _____
Report Sheet 4

V. Answer the following questions:

   A. Where does the photo–phase of photosynthesis occur? _____
      _____

   B. Where does the synthesis–phase of photosynthesis occur? _____
      _____

   C. What events occur during the light–energy conversion stage of photosynthesis?
      _____
      _____
      _____
      _____

   D. What events occur during the carbon dioxide conversion stage of photosynthesis? Include the reactants and products of this stage.
      _____
      _____
      _____
      _____
      _____

   E. List the uses of the products of photosynthesis. _____
      _____
      _____
      _____

Photosynthesis Student ID # _____
Report Sheet 5

VI. Compare and contrast photosynthesis with cellular respiration.

	Photosynthesis	Cellular Respiration
Organisms involved	_____	_____
Reactants	_____	_____
Type of energy process	_____	_____
Organelles involved	_____	_____

5   MITOSIS AND MEIOSIS

LESSON OBJECTIVES

Upon completion of this laboratory exercise, the student should be able to:

1. Observe the stages of mitosis in plant and animal material using the microslide viewer sets.

2. Draw, label and explain the stages of mitosis.

3. Draw, label and explain the phases of meiosis.

4. Differentiate between the First Meiotic Division and the Second Meiotic Division, defining each phase.

5. Explain the significance of meiosis as related to human reproduction.

MATERIALS NEEDED

1. *Photo Atlas for Biology*
2. Pencils—2 colors

PREPARATION

Read the discussion which follows carefully before attempting to complete the exercise. Also read the appropriate chapter(s) in your textbook.

DISCUSSION

Mitosis

The process whereby animals and plants grow and/or repair injured cells is a type of cell division termed **mitosis**. By definition, mitosis is the process during which the nuclear material duplicates and divides. The division of the nuclear material is termed **karyokinesis** and the later division of the cytoplasm is termed **cytokinesis.** The mitotic process begins with one parent cell and culminates in two daughter cells containing the same number and kinds of chromosomes as the original parent cell.

Undifferentiated cells grow and divide rather freely. For example, the epithelial cells lining the intestinal walls and of the skin may undergo division every few days or every few hours in some instances. More specialized cells grow much slower and divide less frequently and some lose the ability to divide altogether. Mature nerve cells never divide; therefore, once destroyed by injury they cannot be replaced. Red blood cells do not contain a nucleus when mature; therefore, mature red blood cells cannot divide.

Each plant or animal species has a characteristic number of chromosomes. The **somatic cells** (body cells) of humans contain 46 chromosomes. These cells are said to contain the **diploid** number (2N) of chromosomes. **Gametes** or sex cells of humans contain one–half the normal chromosome compliment or 23 chromosomes. These cells are said to contain the **haploid** (N) number of chromosomes.

Mitosis is an on–going process in which the changes occur without interruption. Biologists have conveniently identified four phases in the mitotic process in order to explain the events more clearly. Prior to mitosis, the cell grows to its maximum or "adult" size. This stage of the cell's life cycle is termed **interphase**. The nucleus, chromatin material, nuclear membrane and one or more nucleoli are distinct to the observer. Small bodies termed centrioles are visible outside the nucleus in animal cells; these are not present in plant cells.

The first stage of mitosis, termed **prophase**, is characterized by the appearance of double-stranded chromosomes as the chromatin material condenses and forms strands or coils containing the hereditary material DNA. The double–stranded chromosomes, termed chromatids, are actually two identical strands attached at some point, termed the centromere. The nucleoli and nuclear membrane disappear leaving the chromosomes suspended in the nucleoplasm and randomly distributed.

In plant cells, spindle fibers begin to grow toward each other after the chromosomes have appeared. In animals, the centrioles move apart toward opposite sides of the cell producing short fibers in a star–shaped fashion termed an **aster**. The short fibers lengthen into long spindle fibers. In both plant and animal cells, the spindle fibers will reach across the cell to meet other spindle fibers near the equator of the cell. The double–stranded chromosomes or chromatids have migrated toward the equator or midline of the cell.

Stage two of mitosis, **metaphase**, is distinguished as the phase during which the chromosomes become attached to the spindle fibers at the centromeres. When the chromosomes have become attached to the fibers, **anaphase** begins. The chromatids split at the centromeres and move toward opposite ends of the cells. The chromatids are now termed chromosomes once again.

The last phase of mitosis is **telophase**. The chromosomes are clustered at each pole where a nuclear membrane forms around them. The nucleoli reappear and the spindle fibers disappear. In plant cells a cell plate forms between the newly assembled nuclei. The animal cell membrane pinches inward between the newly assembled nuclei in what is termed a cleavage furrow.

The resulting new daughter cells now are in interphase. The chromosomes will diffuse and form chromatin material. The cell will complete its structures and organelles and continue to grow until it reaches its "adult" size at which time the mitotic process will begin again.

Meiosis

Meiosis is a type of cell division that gametes or sex cells undergo. It results in the haploid number of chromosomes for each daughter cell, rather than the diploid or full number of chromosomes that mitosis produces in the daughter cells. Meiosis is referred to as a reduction division because the number of chromosomes in the daughter cells is reduced by one–half. In humans the somatic or body cells contain 46 chromosomes or 23 pairs of chromosomes. The gametes or sex cells contain only 23 chromosomes, one member of each chromosome pair. The necessity for meiosis can readily be seen if you consider the events which occur during human fertilization. If a sperm containing 46 chromosomes united with an egg which also contained 46 chromosomes, the resulting organism would have 92 chromosomes in all of its cells. This is not the case, however, as the human egg contains 23 chromosomes, as does the sperm, resulting in offspring whose somatic cells contain 46 chromosomes. Thus, a constant chromosome number is maintained from generation to generation.

For convenience in studying meiosis, biologists divide the process of meiosis into two major divisions known as the **First Meiotic Division** and the **Second Meiotic Division** with each division being further divided into phases. Remember that meiosis is a continuous event as was mitosis and the descriptive phases only make our study easier.

Prior to the beginning of meiosis, the cell is in **Interphase**. The hereditary material in the nucleus is replicated but the chromosomes are not yet visible. The first meiotic division begins with **Prophase I**. During prophase I the chromosomes appear first as elongated filaments and pairing of **homologous chromosomes** occurs. This pairing of homologous chromosomes (chromosomes which are morphologically alike but not necessarily identical genetically) is termed **synapsis** which results in a synaptic pair of chromosomes or a **tetrad**.

At this point in meiosis each chromosome consists of 2 chromatids united by a centromere and thus they lie very close to each other. A swapping or exchange of equal amounts of hereditary material in chromosomes can occur at this time. This common event is known as crossing-over. The chromosomes shorten and thicken, the nuclear membrane disappears, and the spindle fibers appear. In late prophase the tetrads attach to a spindle fiber and begin moving toward the equator of the cell.

Metaphase I is characterized by the tetrads being lined up at the equator of the cell. **Anaphase I** follows metaphase as in mitosis; however, the centromeres do not split as they do in mitosis. Double-stranded chromosomes migrate toward opposite poles and the cell membrane begins to pinch in or invaginate in the area of the equatorial plane. In other words, the homologous chromosomes (the chromosome pairs) split with one member of each chromosome pair going to opposite ends of the cell.

Telophase I is characterized by the reappearance of the nuclear membrane around the double-stranded chromosomes and complete division of the cytoplasm producing two cells. The two resulting cells enter the phase known as **Interkinesis** during which the cell reorganizes and prepares for the **Second Meiotic Division**. Replication does not occur during interkinesis.

Double-stranded chromosomes (chromosomes consisting of two chromatids) reappear attached by centromeres and the nuclear membrane disappears in **Prophase II**. In **Metaphase II** the double-stranded chromosomes line up along the equatorial plane and attach to spindle fibers. The centromeres divide and the chromatids of the chromosomes are pulled toward opposite poles by the spindle fibers as the cell membrane begins to pinch in the region of the equatorial planes as **Anaphase II** proceeds.

In **Telophase II** the nuclear membrane begins to reappear and further invagination of the cell membrane divides the two cells which entered the Second Meiotic Division once again, resulting in four haploid cells.

Meiosis in males is termed **spermatogenesis** and produces four new cells or sperm equal in size. In females, only one full-size ovum is produced along with three small cells termed **polar**

bodies. The ovum receives most of the cytoplasm during **oogenesis** (the term for meiosis in females). The polar bodies very quickly disintegrate and are reabsorbed by the body.

Meiosis results in cells that contain only one member of each chromosome pair. Mitosis results in cells that have the full chromosome complement. Because the gametes undergo meiosis, with the resulting reduction in chromosome number and because the separation of the chromosome pairs is strictly random, different offspring produced by two individuals will have different characteristics. In other words, the children in a family will all look different yet will share certain family characteristics.

LABORATORY EXERCISE PROCEDURE

Carefully read all of the discussion material before attempting the following exercises. Carefully complete each exercise recording the required information on the report sheet. When complete, submit the reports to the instructor for grading.

LABORATORY EXERCISES

I. Mitosis

    A. Look at pages 10–13 in the *Photo Atlas* and observe figures 10a–13f. Observe the five mitotic phases in each set. Space has been provided for sketching the phases occurring in both plant and animal cells. Compete the drawings with as much detail as possible and label all structures and phases. There will be at least 10 drawings (14 drawings if 2 daughter cells for the plant and 2 daughter cells for the animal are included). Complete the drawings on Report Sheet 1.

    B. Complete the questions on Report Sheet 2.

    C. Complete the questions on Report Sheet 3.

II. Meiosis

    A. In Part IV on the report sheet, space has been provided for use in performing the following exercise.

    The problem will involve an organism, the fruit fly *Drosophila melanogaster,* which has only four pairs of chromosomes in its somatic cells including the spermatogonia.

    The question is sometimes asked "Why the fruit fly again?" and perhaps a brief explanation as to why so many experiments have been done using this tiny organism is in order. *Drosophila melanogaster* is very small, therefore, it can be raised in small vials or bottles if provided with easily obtained food supplies such as banana, agar and yeast. The entire life cycle is complete in 10 days to 2 weeks with the females laying large numbers of eggs. It has 4 pairs of relatively large chromosomes in the cells of its salivary glands, thus it is easier to study than a human cell containing 46 chromosomes.

    The chromosomes are numbered Pair I, II, III, and IV. Pair II, III, and IV are identical in both male and female fruit flies. Pair I, termed the sex chromosomes, are alike in the female and these are called the X chromosomes. In the male there is one X chromosome and a Y chromosome with a hooked end. The sperm thus determines the sex of the offspring as is true in humans.

The problem is to start with a fruit fly spermatogonia in the interphase and take it through the entire meiotic process. Label the important structures in each phase and include a brief explanation of the events occurring in each phase. Use the following labels as needed:

Interphase	Telophase I
Nucleus	Interkinesis
Centrosomes	Prophase II
Spindle fibers	Metaphase II
Double–stranded chromosome	Anaphase II
Metaphase I	Telophase II
Anaphase I	4 sperm

Remember that chromosomes are found in pairs and that one chromosome of each pair is provided by the female parent and one is provided by the male parent. Use 2 colors or pen or pencil to distinguish between the two. Choose four shapes such as a long rod, a short rod, a circle, and a bent rod to distinguish between the four pairs of chromosomes. If problems arise, consult the textbook or review the discussion in this lab manual. Start at the top of Report Sheet 4 with the following hypothetical cell:

FIGURE 1

B. Answer the questions on the report sheet.

Mitosis and Meiosis

MITOSIS AND MEIOSIS

Report Sheet 1

Name _____

Student ID # _____

Campus _____

Date _____

I. Mitosis   (Make sketches from pages 10–13 in the *Photo Atlas for Biology*)

        Animal   Plant

Mitosis and Meiosis Student ID # _____
Report Sheet 2

B. Complete the following sentences using the terms provided and place the letter of the answer in the blank proceeding the number of the sentence.

___ 1. The _____ cells of humans contain _____ chromosomes.

___ 2. Human somatic cells contain the _____ number of chromosomes.

___ 3. The process during which the nuclear material duplicates and divides is called _____ .

___ 4. Functions of mitosis include _____ and _____ .

___ 5. Once destroyed, mature _____ cells can never be replaced because they never divide.

___ 6. The _____ cells of the skin may undergo division every few hours.

___ 7. _____ is the term used for the division of the nuclear material.

___ 8. The division of the cytoplasm is termed _____ .

___ 9. Human sex cells (gametes) contain one–half the normal chromosome complement or _____ chromosomes. This is the _____ condition.

A. cytokinesis
B. diploid
C. epithelial
D. 46
E. growth
F. haploid
G. karyokinesis
H. mitosis
I. nerve
J. repair of injured cells
K. somatic
L. 23

Mitosis and Meiosis Student ID # _____
Report Sheet 3

C. Answer the following questions: (answers can be found in the textbook and the discussion part of this exercise or refer to pages 10–13 in the *Photo Atlas*).

What difference, if any, exists between plants and animals during mitosis?

	Animal	Plant
Prophase	_____	_____
	_____	_____
Metaphase	_____	_____
	_____	_____
Anaphase	_____	_____
	_____	_____
Telophase	_____	_____
	_____	_____
Interphase	_____	_____
	_____	_____

D. Briefly describe the events that occur during:

Prophase _____

_____

Metaphase _____

_____

Anaphase _____

_____

Telophase _____

_____

Interphase _____

_____

Mitosis and Meiosis
Report Sheet 4

Student ID # _____

I. A. Meiosis

Mitosis and Meiosis Student ID # _____
Report Sheet 5

1. What is the diploid number (2N) of chromosomes in the fruit fly? _____

   The haploid number (N)? _____

2. Why is the haploid condition desirable in gametes or sex cells?

   _____

   _____

3. Does spermatogenesis or oogenesis produce the greater number of viable cells?

   _____

6 INTRODUCTION TO GENETICS

LESSON OBJECTIVES

Upon completion of this laboratory exercise the student will be able to:

1. Examine Gregor Mendel's principle of segregation and principle of independent assortment.

2. Become familiar with basic genetic vocabulary and notation.

3. Work genetics problems involving both monohybrid and dihybrid crosses.

MATERIALS NEEDED

Laboratory manual
Pencil

PREPARATION

Read the discussion which follows carefully before attempting to complete the exercise. Also read the appropriate chapter in your textbook.

DISCUSSION

The science of genetics has greatly advanced since the time of Gregor Mendel. He was the first to systematically study inheritance by means of controlled experiments using the common garden pea (*Pisum sativum*). Mendel's studies lasted eight years and resulted in several basic concepts including those of dominant and recessive factors, the Principle of Segregation and Mendel's Theory of Independent Assortment.

Mendel observed that some traits or characteristics are always expressed if the same traits or characteristics were present in the parent generation, while others show up only some of the time. When he crossed a tall pea plant and a short pea plant all of the resulting plants were tall. However, when two of the resulting tall plants were crossed, a few short plants appeared. Mendel applied the term dominant to the expressed trait and recessive to the factor not expressed.

In his experiments with pea plants, Mendel determined that the trait for tallness was dominant over shortness. Mendel assumed that each parent plant contributed only one allelic form of a gene to their offspring or progeny. In his Principle of Segregation, he theorized that during meiosis members of a homologous pair of chromosomes separated, each going into the formation of separate gametes. Regarding the height of the pea plants, Mendel recognized a gene for tallness, T; and a gene for shortness, t. These genes, T and t are termed alleles

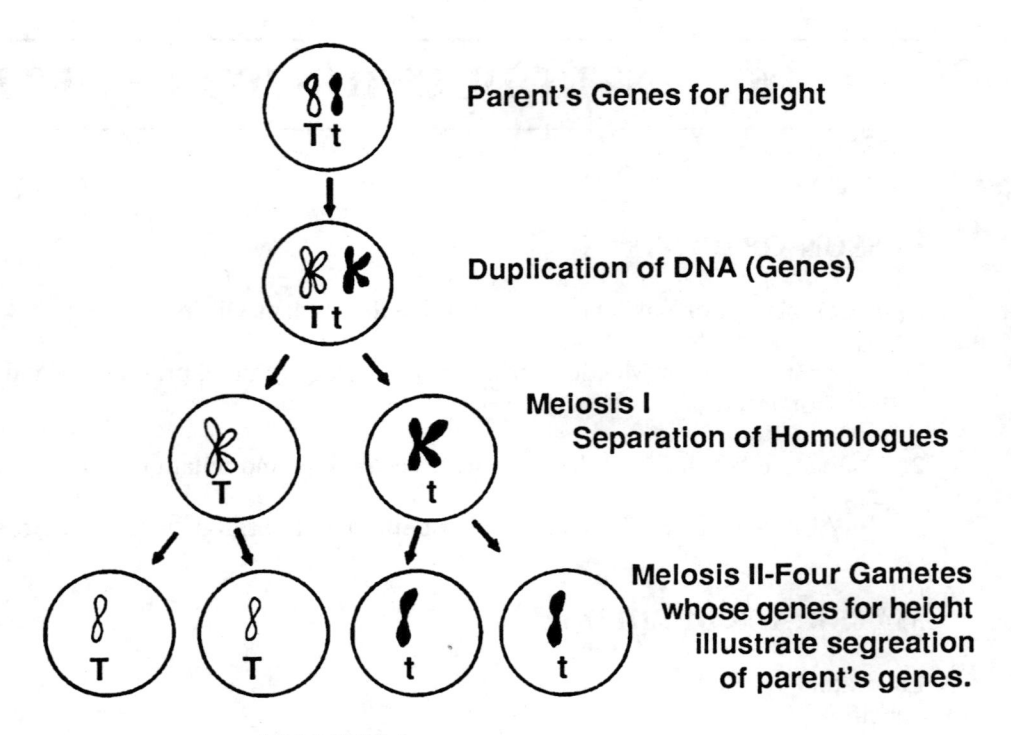

FIGURE 1
Mendel's Principle of Segregation

Parent's Genes for height — Tt

Duplication of DNA (Genes) — Tt

Meiosis I — Separation of Homologues — T / t

Meiosis II - Four Gametes whose genes for height illustrate segreation of parent's genes. — T, T, t, t

Metaphase I - Showing Possible gene orientation

Left orientation:
- TT tt / RR rr
  - T T / R R → T R, T R
  - t t / r r → t r, t r

Gametes resulting from the above orientation

Right orientation:
- tt TT / RR rr
  - t t / R R → t R, t R
  - T T / r r → T r, T r

Gametes resulting from the above orientation

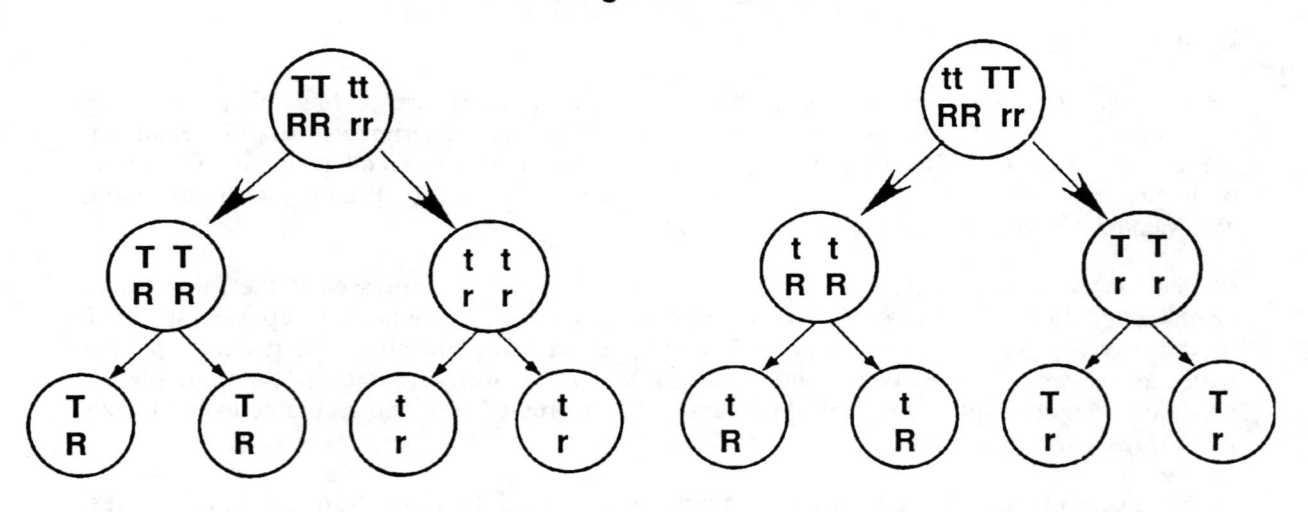

FIGURE 2
Theory of Independent Assortment

50 Exercise 6

because they are contrasting genes found at the same place or locus on homologous chromosomes. See Figure 1.

Mendel's Theory of Independent Assortment states that each allelic pair of genes segregates or separates independently of the allelic pairs for any other trait. This is best represented by choosing two traits such as height of pea plant, T (tall) and t (short); and color of pea flower, R (red) and r (white). During metaphase the homologous chromosomes orientate themselves randomly and independently of the other homologous pair. Refer to Figure 2.

The physical features of an animal or plant which can be measured and observed are referred to as the **phenotype** of the organism. For example, if you have blue eyes, your phenotype for eye color is blue. The **genotype** refers to the genetic make-up of an organism and is expressed as alphabetical letters. The blue-eyed person has a genotype of bb. Capital letters are used to represent dominant alleles or traits and small or lower case letters represent the recessive alleles, thus, blue eyes are the recessive condition.

The genotype of an individual can be either homozygous (like) or heterozygous (unlike). Homozygous organisms have like genes for the same trait; and therefore, can produce only one type of gamete. The blue-eyed individual has a genotype of bb-like genes—so this individual is homozygous recessive for eye color. We know he or she is recessive because the alphabetical letters are small or lower case letters.

An organism is heterozygous if it has unlike genes for the same trait. Heterozygous organisms can produce two types of gametes. An individual with the genotype Bb for eye color has the phenotype of brown eyes because the dominant gene (B) is expressed over the recessive gene (b). This individual would produce two types of gametes, one containing the gene B; the other containing the gene b.

To solve any genetics problem, it is necessary to determine all the various types of gametes that each parent can contribute to the offspring. In genetic crossings, a monohybrid cross involves only one trait. An example of a monohybrid cross would be the cross of a tall pea plant (TT) with a short pea plant (tt). These parent plants would be the parental generation or the P_1 generation. The tall plant can contribute only T gametes; the short plant can contribute only t gametes. The resulting plants from such a cross are the filial generation or the F_1 generation. An F_2 generation would be the result of a cross between F_1 progeny.

LABORATORY EXERCISE PROCEDURE

To determine if the basic genetic principles discussed are understood, do the following genetic crosses and answer the appropriate questions. When the worksheets are complete, *transfer the answers to the Report Sheet* and submit *only the report sheet* to the instructor for grading. Credit will not be given for the work sheets. For help with the genetics problems, carefully read the discussion part of this exercise and the appropriate chapter or chapters in your textbook.

LABORATORY EXERCISES

    Example Problem: Cross a homozygous tall plant (TT) with a homozygous short plant (tt).

    a. List the possible gametes: T, T, t, t

Introduction to Genetics   51

b. Using the Punnet square, combine the gametes.

	T	T
T	Th	Th
h	Th	Th

c. What are the phenotypes of the F_1 plants? _____all tall_____

d. What is the genotypes of the F_1 plants? _____Tt_____

I. Cross two F_1 individuals from problem 1 to produce the F_2 generation using the Punnett square provided.

a. What are the possible gametes from each F_1 plant? _____ and _____ ; _____ and _____ .

b.

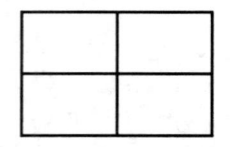

c. How many tall plants are produced? _____ How many short plants? _____ (This is known as the phenotypical ratio and is written as 3 tall: 1 short.)

d. How many plants are:

homozygous dominant, TT _____

heterozygous tall, Tt _____

homozygous recessive, tt _____

(This is termed the genotypical ratio and is written 1:2:1).

II. Mendel also studied the texture of seed coats in his pea plant experiments. He determined that smooth seed coats (S) were dominant over wrinkled (s) seed coats. Cross a homozygous dominant smooth seed with a homozygous recessive wrinkled seed.

a. P_1 individuals genotypes _____ × _____

b. Possible gametes _____ , _____   _____ , _____

52   Exercise 6

c. Using the Punnet square, complete the cross.

d. What are the genotypes of all of the F_1 individuals? _____

e. What is the phenotype of all of the F_1 individuals? _____

III. Now cross the two plants from the F_1 generation above.

a. F_1 individuals genotypes _____ × _____

b. Possible gametes _____ , _____ and _____ , _____

c. Using the Punnet square, complete the cross.

d. What is the phenotypic ratio? _____

e. What is the genotypic ratio? _____

The crossings done so far are all monohybrid crosses. A dihybrid cross involves two separate traits whose genes are located on two different pairs of homologous chromosomes. In addition to the texture of the seed coat, Mendel studied the color of the seed pod (Yellow, Y and green, y).

IV. Cross a homozygous dominant smooth, yellow seeded plant with a homozygous ecessive wrinkled, green seeded plant.

a. The P_1 genotypes are _____ × _____

b. The possible gametes are _____ , _____ , _____ , and _____ .

c. Using the Punnet square, complete the dihybrid cross.

d. What are the phenotypes of all of the F_1 individuals? _____

e. The genotype? _____

V. You are a helper in Mendel's pea garden. Mendel has given you a sack of pea seeds and told you to count the smooth seed coats and wrinkled seed coats. You have counted 690 smooth (S) and 230 (s). These are the F_2 plants.

a. The phenotypic ratio is _____ : _____

b. The P_1 genotypes are _____ × _____

c. What is the genotype of the F_1 generation? _____

d. Cross two of the F_1 plants to see if your answer is correct.

VI. In problem V a homozygous dominant smooth yellow seed plant was crossed with a homozygous recessive wrinkled, green seed plant. The genotype of the F_1 plants was SsYy; all the F_1 plants were heterozygous smooth, yellow seeded plants. Cross two plants from the F_1 generation in the Punnet square provided below.

a. What are the possible gametes? _____, _____, _____, _____ and

_____, _____, _____, _____.

b. The plants represented in the Punnet square above are termed the F_1, F_2 (circle one) generation.

c. What are the phenotypes of the F_2 plants?

_____ – _____ (How many?)

_____ – _____ (How many?)

_____ – _____ (How many?)

_____ – _____ (How many?)

In the dihybrid cross above there are sixteen possible combinations. If VI b and VI c are answered correctly, it is noted that the possible phenotypes obtained in a dihybrid cross form a 9:3:3:1 ration. Mendel assumed that the possible phenotypic genetic combinations in a mating or cross were governed by the laws of chance and probability, and research has confirmed that this is so. The practical usefulness of the science of genetics is its ability to predict the phenotypes and genotypes of animal and plant progeny.

VII. The ability to taste the chemical phenylthiocarbamide (PTC) is due to a dominant ene (T) in man. The inability to taste PTC is due to its recessive allele (t). A man who can taste PTC, but whose mother could not, marries a woman who can taste PTC, but whose father could not.

a. What is the genotype of the man? _____ Explain _____

_____

_____

b. What is his wife's genotype? _____ Explain _____

_____

_____

c. What types of gametes will the man produce? _____

What types of gametes will the woman produce? _____

d. What proportion of their children will probably have the ability to taste PTC? (Hint: draw a Punnet square and do a test cross.)

_____ : _____

Introduction to Genetics 55

VIIII. In humans, migraine (a type of headache), is due to a dominant gene (M). A woman suffers from migraine which her mother also had. She has a sister and a brother; neither suffers from migraine, nor did the father.

What is the genotype of:

a. Her father _____

b. Her mother _____

c. The woman _____

d. If she marries a man who is heterozygous (Mm), what is the probability that their offspring will suffer from migraine?

_____

INTRODUCTION TO GENETICS

Report Sheet 1

Name _____

Student ID # _____

Campus _____

Date _____

I. a. Possible gametes from each F₁ plant: _____ and _____ ; _____ and _____ .

   b.

   c. Number of tall plants produced? _____ .

   Number of short plants produced? _____ .

   d. How many plants are

   homozygous dominant, TT _____

   heterozygous, Tt _____

   homozygous recessive, tt _____

II. a. P₁ individual's genotypes _____ × _____

   b. Possible gametes _____ _____

   c.

   d. Genotype of F₁ _____

   e. Phenotype of F₁ _____

Introduction to Genetics   57

Introduction to Genetics Student ID # _____
Report Sheet 2

III. a. F_1 genotypes _____ × _____

    b. Possible gametes _____ and _____ ; _____ and _____ .

    c.

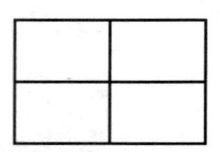

    d. What is the phenotypic ratio? _____

    e. What is the genotypic ratio? _____

IV. a. The P_1 genotypes are _____ × _____

    b. Possible gametes are _____ , _____ , _____ , and _____ .

    c.

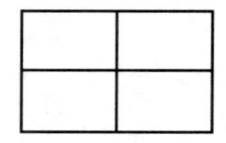

    d. Phenotypes of the F_1 plants? _____

    e. Genotypes of the F_1 plants? _____

V. a. The phenotypic ratio is _____ : _____

    b. The P_1 genotypes are _____ × _____

    c. What are the genotypes of the F_1 generation? _____

    d.

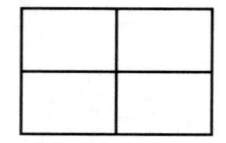

Introduction to Genetics Student ID # _____
Report Sheet 3

VI. a. Possible gametes _____, _____, _____, _____ and _____, _____, _____, _____.

 b.

 c. Phenotypes of the F_2 plants?

   smooth yellow = _____ (how many?)

   smooth green = _____

   wrinkled yellow = _____

   wrinkled green = _____

VII. a. Genotype of the man _____ Explanation _____
_____

 b. His wife's genotype _____ Explanation _____
_____

 c. Gametes produced by the man _____

   Gametes produced by the woman _____

 d. _____ : _____

VIII. a. Father's genotype _____

 b. Mother's genotype _____

 c. Woman's genotype _____

 d. _____ : _____

7 HUMAN GENETICS

LESSON OBJECTIVES

Upon completion of this laboratory exercise the student will be able to:

1. Learn some of the inherited human traits

2. Determine individual phenotype and possible genotype for several common traits.

3. Construct a pedigree chart showing how a common human trait is transmitted from one generation to the next.

MATERIALS NEEDED

1. Laboratory Manual
2. Pencil
3. PTC paper*

   * Item supplied in lab kit

PREPARATION

The genetic terms defined in the Laboratory Exercise "Introduction to Genetics" will be used in this exercise without repeating previous discussions of their meanings. If necessary, review those materials before proceeding with Human Genetics. Read the appropriate chapter(s) in your textbook as well.

DISCUSSION

The human gene pool is the result of more than 4 billion years of organic evolution. Our DNA contains the codes which make us biologically different from all other life forms. We have only to observe members of our own families, friends and acquaintances to see the vast numbers of variations and the remarkable similarities which occur.

For these reasons, human genetics has a popular appeal to many students who may wonder about differences in hair color, blood types or height. Unfortunately, *Homo sapiens* (humans) are not suitable subjects for genetics studies because their life span is long, families are small, and test matings are prevented by laws and ethics. However, some rather superficial traits can be noted and this laboratory exercise examines these.

LABORATORY EXERCISE PROCEDURE

In this lab exercise, you will determine your phenotype and possible genotype for the indicated traits. If the trait is the result of a dominant gene, you will record a single capital letter designating the known allele and a dash (–) for the unknown allele. Recessive traits which you possess will be recorded by two small letters because you must be homozygous genotypically for the trait to be expressed phenotypically.

Illustrations for hair whorl and pointer's and hitchhiker's thumb are found in the lab manual. Some other genetic traits are described in the lab manual.

Facial dimples are dominant but may be inherited as an irregular dominant trait. Absence of facial dimples generally indicates a homozygous recessive genotype. No record of this is required on the report sheet.

LABORATORY EXERCISES

I. Superficial Inherited Traits

   A. Sex is perhaps the easiest trait for which to determine your phenotype because you are either male or female. In *Homo sapiens*, sex is determined by a pair of chromosomes. Males have one long chromosome designated "X" and a shorter chromosome designated "Y". Females have two long "X" chromosomes. Record your phenotype and genotype on the Report Sheet.

   B. Hair whorl direction is another inherited characteristics. The hair at the back of the head should be observed. The dominant condition is hair at the back of the head whorling in a clockwise direction. The recessive condition is hair whorls in a anti– or counter–clock– wise direction. Use W for the dominant condition; for counter–clockwise whorling use w.

   C. Hyperextensibility of the thumb joint is also a dominantly inherited trait. Two conditions exist which are of interest. The ability to position the thumb at a right angle (sometimes known as pointer's thumb) will be designated by HH or Hh (H–). The inability will be the homozygous recessive hh. Hyperextension of the thumb back toward the hair is alsoa dominantly inherited trait sometimes known as hitchhiker's thumb. The genotypes for hitchhiker's thumb are HH or Hh (H–); for non–hitchhiker thumb the genotype will be hh.

   D. Examine your ear lobes in a mirror. Ear lobe attachment is termed "free" if the lower part of the ear lobe is unattached to the side of the head or "attached" if the lower part of the ear lobe is attached to the side of the head. The dominant unattached or free ear lobes are indicated by LL or L–. The person with attached ear lobes has the genotype ll which is a homozygous recessive combination. Record your genotype for ear lobe attachment.

   E. The presence of hair on the middle phalanx of the finger is due to the presence of a dominant allele which can be symbolized by P. The fact that mid–phalangeal or mid–digital hair is a dominant trait means that the genotype for this trait could be either PP or Pp. Record your genotype as P– if you possess mid–phalangeal hair. (If you knew the genotype of your parents for this trait you could be certain of the missing allele.) Absence of hair from skin covering the middle bones of your fingers is due to a recessive gene in the homozygous condition "p". In either case record your genotype and pheno- type for this trait on the Report Sheet.

F. The ability to taste PTC (phenylthiocarbamide) paper is not possessed by all humans. The taste of PTC paper is extremely bitter to tasters who are genotypically TT or Tt. Non-tasters are homozygous recessive tt. You were given a strip of PTC paper in your lab kit. Chew it up; working it toward the back of your tongue. **CAUTION: DON'T SWALLOW THE PAPER. THE CHEMICAL IS HARMLESS; BUT THE HUMAN DIGESTIVE SYSTEM IS NOT DESIGNED TO DIGEST PAPER.** If the paper tastes bitter you are a taster (T-). If it tastes like plain tissue paper, you are a non- taster (tt). Record your results.

G. To determine the presence of the dominantly inherited widow's peak hairline, return to your mirror once again. If the hair recedes from the widow's peak (a point over the forehead) your genotype will be recorded as W-. If your hairline is straight across your head your genotype is ww. Record your phenotype as being either Widow's Peak or Straight Hairline.

H. The ability to roll the sides of the tongue up to form an U shape is determined by a dominant gene. The symbol used is R- for the tongue rollers and rr for the non-rollers. Again, look in the mirror and determine your genotype and phenotype for tongue rolling and record these on the Report Sheet.

I. Another dominant gene symbolized here by BB or Bb determines whether the last joint of the little finger bends toward the other fingers. Straight joints on the little finger indicates the presence of a homozygous recessive condition "bb". Position your hands flat on a flat surface and determine your phenotype and genotype for this trait. Record your findings on the Report Sheet.

J. Length of Big Toe – The relative lengths of the big toe and second toe are easily determined inherited traits. Those individuals possessing a long big toe (hallux) are genotypically recessive (ll) for the trait. Possession of the dominant gene results in a short big toe in relation to the second toe. Sometimes, but rarely, the phenotype may differ on each foot. If one or both big toes are either equal to or greater in length than the second toe, the individual has the recessive gene. Record your genotype (either L- or ll) and phenotype on the Report Sheet.

K. Some traits are termed sex-influenced. Sex-influenced alleles are those whose dominance is affected or altered by the sex of the individual. The relative lengths of the index (pointer) and ring fingers is such. The condition of a short index finger in relation to the fourth or ring finger is dominant (F-) in males and recessive (ff) in females. Sometimes variations occur between the right and left hands. If one or both index fingers are equal to or greater in length than the fourth finger, the recessive genotype is present in males and the dominant genotype is present in females. Record your genotype and phenotype on the Report Sheet.

II. Try to establish a pattern of inheritance for *one* of the following traits which you have examined: ear lobe attachment, tongue rolling, thumb joint hyperextensibility, hairline, mid-digital hair, or bent-joint little finger. Seek cooperation of your parents, grandparents, brothers, sisters, and your own children (if you have offspring). If your relatives live a long distance, you might choose hairline and use photographs to determine an inheritance pattern. Include as many individuals as you can to make your chart more meaningful to you.

Space is provided on the Report Sheet for the construction of your pedigree chart. A circle usually indicates a female; a square usually indicates a male offspring. The following pedigree chart for the inheritance of the ability to taste PTC in the author's family is presented as an example. Remember, the ability to taste is dominant; therefore, a non-taster has a homozygous recessive tt condition.

FIGURE 1
Pedigree Chart for the Inheritance of the Ability to Taste PTC

○ = female, taster

□ = male, taster

● = female, non-taster

■ = male, non-taster

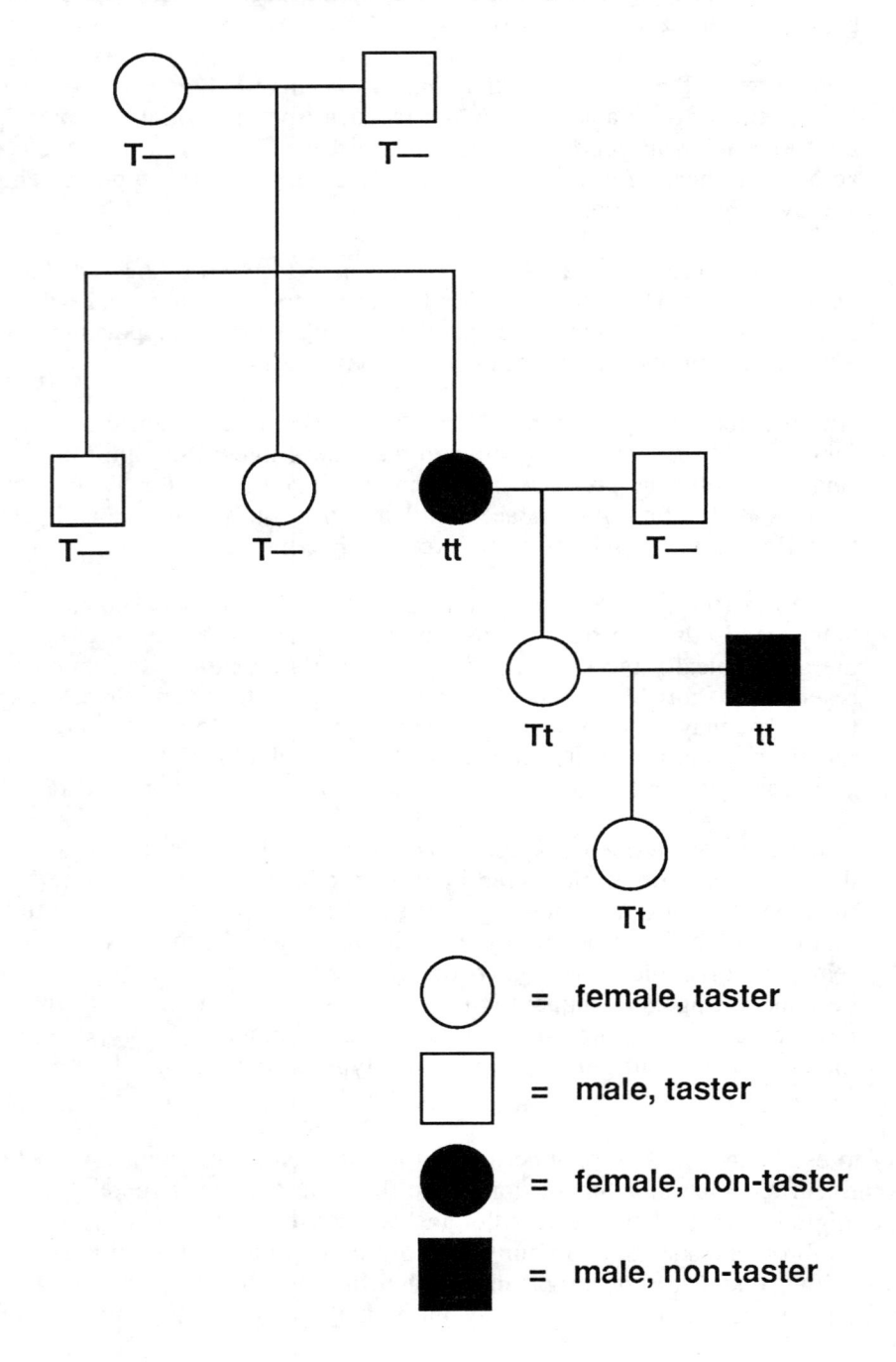

64   Exercise 7

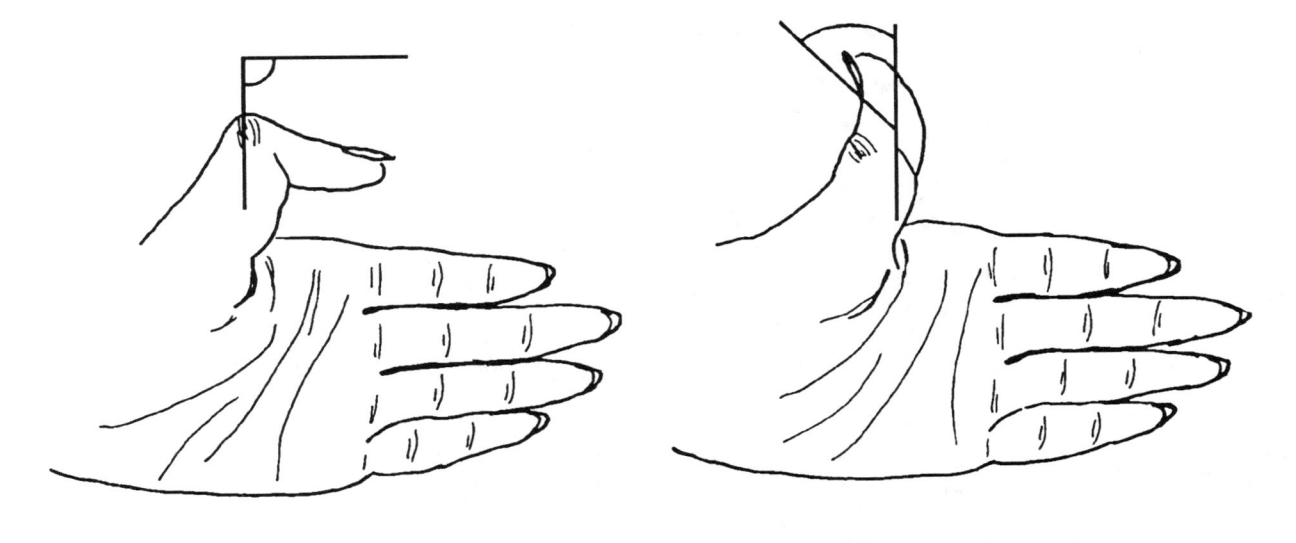

POINTER'S THUMB

HITCHHIKER'S THUMB

FIGURE 2

Dominant

Recessive

FIGURE 3

Human Genetics   **65**

HUMAN GENETICS

Report Sheet 1

Name _____

Student ID # _____

Campus _____

Date _____

I. Superficial Inherited Traits

		Your Phenotype	Your Genotype
A.	Sex	_____	_____
B.	Hair whorl direction	_____	_____
C.	Thumb joint hyperextensibility		
	Right Angle	_____	_____
	Hitchhiker's Thumb	_____	_____
D.	Ear lobe attachment	_____	_____
E.	Mid–digital hair	_____	_____
F.	PTC tasting	_____	_____
G.	Hairline		
	Widow's peak OR	_____	_____
	Straight Hairline	_____	_____
H.	Tongue rolling	_____	_____
I.	Bent–joint little finger	_____	_____
J.	Length of big toe	_____	_____
K.	Finger length		
	Index finger	_____	_____
	Ring finger	_____	_____

Human Genetics Student ID # _____
Report Sheet 2

II. Pedigree Analysis of _____
(trait you have chosen)

Key to Pedigree Chart: (Establish your own key or use the one in the example)

8 DNA REPLICATION

LESSON OBJECTIVES

With the completion of this lab the successful student will be able to:

1. Describe the components of a nucleotide.

2. List the bases which appear in DNA and state the base pair relationships.

3. Correctly give the sequence of bases on one-half of a DNA molecule when given one-half of a DNA strand.

4. Explain how DNA replication occurs.

5. Identify the effects of a point mutation during replication.

MATERIALS NEEDED

1. Laboratory Manual
2. Textbook
3. Pen or pencil
4. Construction paper

PREPARATION

Read the discussion which follows carefully before attempting to complete the exercise. Also read the appropriate chapter(s) in your textbook.

DISCUSSION

Deoxyribonucleic acid (DNA) is the chemical means whereby an organism's genetic information is maintained and transmitted from generation to generation by the genes. In addition to this very important function, DNA dictates the structure and controls the function of every cell making up an organism.

Genes are found on chromosomes in the nucleus of a cell. The genes regulate the production of proteins within the cell. The proteins are composed of polymers of amino acids. Polymers are long chains of amino acids. To understand how these enormous feats are accomplished by DNA, it is necessary to examine nucleotides, the structural units which combine to form DNA.

A nucleotide consists of three components: a phosphate group (PO_4), a five-carbon sugar (either ribose or deoxyribose) and a nitrogen-containing compound. The nitrogen-containing compound may be a purine or a pyrimidine. If it is a purine, then it is either adenine or guanine. If it is a pyrimidine, then it is cytosine, thymine, or uracil.

The DNA molecule is a helically arranged double polymer of nucleotides held together by weak bonds between opposing nitrogen–containing compounds termed base pairs. This structure resembles a twisted ladder. The sides of the ladder consist of alternating sugar and phosphate molecules. The rungs of the ladder are the nitrogenous bases projecting from the sugar molecules. The nitrogen–containing compound **adenine** is always bonded to **thymine** by two hydrogen bonds. **Cytosine** is always bonded to **guanine** by three hydrogen bonds. The nitrogen– containing compound uracil is not present in DNA The twisted structure thus described is termed a double helix and was first illustrated by James D. Watson and Francis Crick in the 1950's.

The duplication of DNA involves several steps. First, the DNA helix is split open exposing the base pairs. The hydrogen bonds break, causing the base pairs to separate. Then free nucleotides from the nucleoplasm begin to attach to exposed base sites. Adenine attaches to thymine; thymine attaches to adenine; cytosine attaches to guanine; guanine attaches to cytosine. This attachment of nucleotides continues until the DNA molecule is duplicated. The free nucleotides are supplied from the cytoplasm from a supply maintained by the metabolic breakdown of carbohydrates, fats, and other materials.

Mutations are changes that occur within a DNA strand. **Point mutations** occur when one base is substituted for another base in a triplet of bases (called a codon). The codons are specific for particular amino acids. A point mutation on a given codon changes the amino acid code.

LABORATORY EXERCISE PROCEDURE

Carefully follow the directions for each of the following procedures. Read each procedure and assemble all needed materials before attempting them. When the procedures are complete submit the completed report sheet to the instructor for grading. Also read the appropriate chapter(s) in your textbook.

LABORATORY EXERCISES

I. DNA Strand Formation

You will be making a short sequence of a human gene that controls the body's production of growth hormone. Researchers call this gene the hGH (human growth hormone) gene. It is actually made up 573 nucleotide pairs, but you will only construct the first 10 base pairs of the gene. The sequence is A–A–G–C–T–T–A–T–G–C. (A = adenine nucleotide; T = thymine nucleotide C = cytosine nucleotide, G = guanine nucleotide)

A. Trace the models of a deoxyribose sugar, phosphate unit, adenine, cytosine, guanine and the thymine units on construction paper using the following key:

deoxyribose = white   guanine = blue
phosphate = black   cytosine = yellow
adenine = red   thymine = green

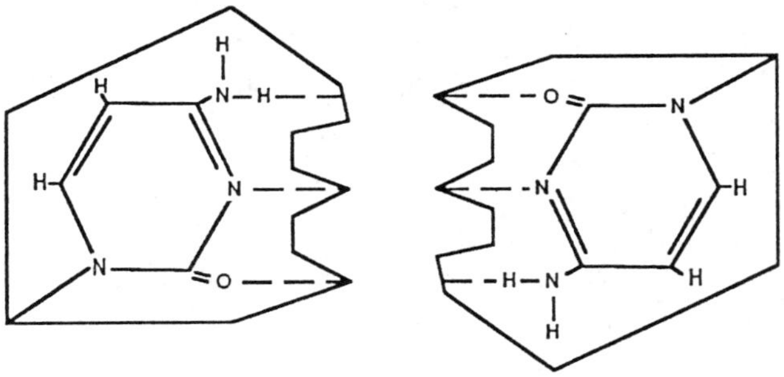

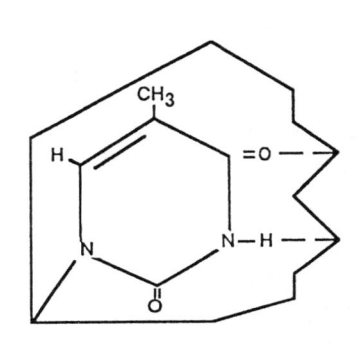

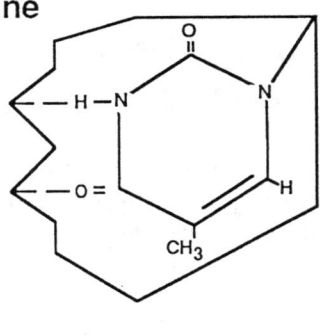

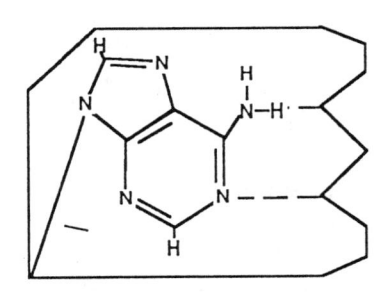

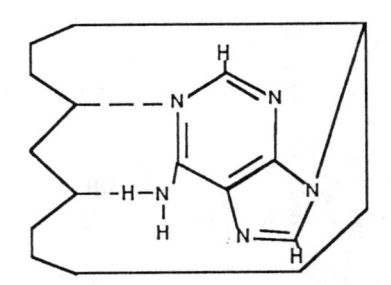

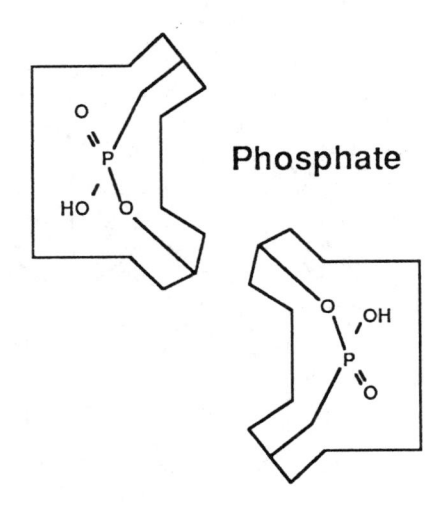

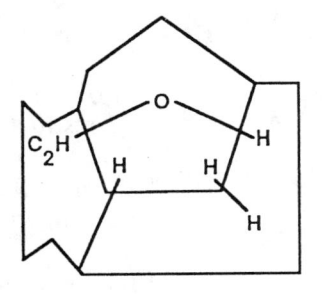

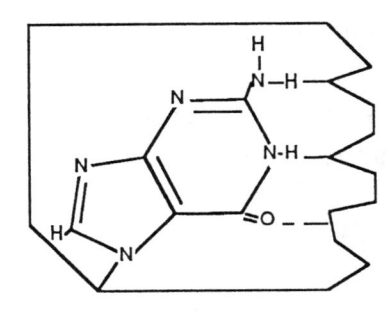

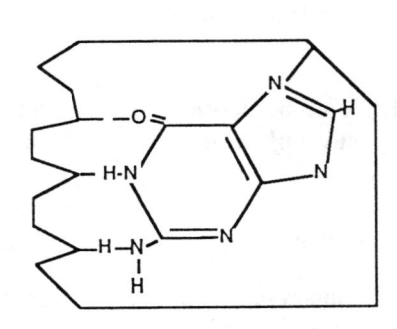

FIGURE 1

DNA Replication 71

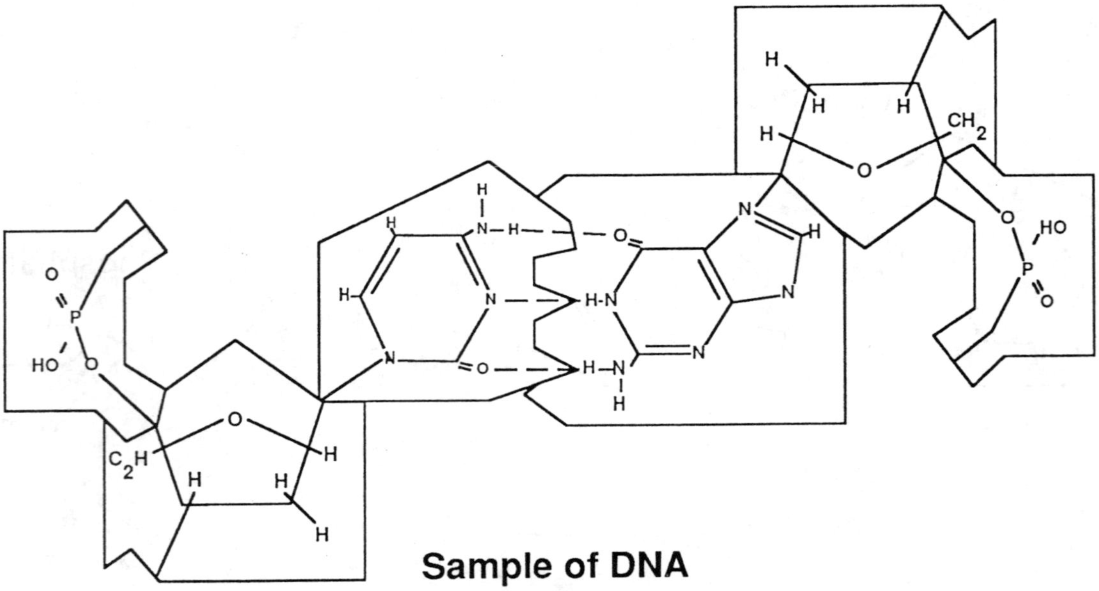

Sample of DNA
FIGURE 2

Cut out enough figures to make the 10 strand hGH DNA molecule segment. One strand of DNA consists of 2 nucleotides. (Hint: If you have cut out 6 Adenine, 6 Thymine, 4 Guanine, 4 Cytosine, 20 phosphate and 20 sugar units, you are proceeding correctly.)

B. Glue the figures on a piece of paper to represent the 10 strand DNA molecule, counting one strand as two phosphate, sugar and nitrogen base combinations (2 nucleotides). Be sure to pay attention to the correct pairing of complementary nitrogen bases.

II. DNA Replication

A. Cut out the figures necessary for another 10 strand segment of the hGH molecule on construction paper.

B. Separate all but two of the strands of nucleotides to demonstrate the zipperlike ashion in which the DNA molecule opens during replication.

C. Link nucleotides to each of the "old" strands of nucleotides. When finished, there should be 2 double-stranded segments. Glue your figures to the paper.

III. Effects of a Point Mutation

A. To demonstrate a point mutation, construct another 10 strand segment of the hGH DNA molecule.

B. Identify the second adenosine nucleotide from the left side of the top strand. Remove this nucleotide and replace it with a cytosine nucleotide. This change represents a point mutation.

C. Glue all pieces to a sheet of paper

IV. Turn in the models with your lab report. Be sure to label each section and identify the point mutation using an arrow pointing to it.

DNA REPLICATION

Report Sheet 1

Name _____

Student ID # _____

Campus _____

Date _____

I. DNA Structure

    A. List the components of a nucleotide _____

    _____

    _____

    B. List the bases in DNA

    _____

    C. In a DNA molecule, which nucleotides normally bond together?

    _____

    _____

    D. How is the bottom strand of your DNA molecule the same as the top strand?

    _____

    _____

    _____

    How is it different? _____

    _____

    _____

II. DNA Replication

    A. Describe the 2 DNA molecules that resulted after replication. _____

    _____

    _____

    Are they similar or different? _____

DNA Replication Student ID # _____
Report Sheet 2

Explain. _____
_____
_____

B. What features of DNA replication cause each new DNA molecule to be exactly like the original?

_____
_____
_____
_____

C. Why is replication an important cell process? _____
_____
_____

D. Will the two DNA molecules produced function the same or differently? _____

Explain. _____
_____
_____
_____

III. Point Mutation

A. How does the point mutation that you made change the DNA code on your molecule?

_____
_____
_____

DNA Replication
Report Sheet 3

Student ID # _____

    B. Referring to your text, explain how mutations can occur in cells and how such changes might affect a new organism with the mutation.

_____

_____

_____

_____

_____

IV. DNA Models

    Be sure to attach your DNA models and submit them this lab report.

PHOSPHATE   GUANINE   CYTOSINE

SUGAR GROUP   THYMINE   ADENINE

SAMPLE OF DNA

9 RNA AND PROTEIN SYNTHESIS

LESSON OBJECTIVES

Upon completion of this laboratory exercise the student will be able to:

1. Compare and contrast DNA and RNA.

2. List the types of RNA and give a function of each type.

3. Describe how amino acids are linked into a protein in the ribosome and the fate of mRNA and tRNA segments after the ribosome has used them.

4. Describe protein synthesis using written or oral explanations and illustrations.

5. Identify a specific incidence of chromosomal mutation resulting in harmful effects.

6. Answer all questions on the lab report sheets.

MATERIALS NEEDED

1. Laboratory manual
2. Textbook
3. Pen or pencil

PREPARATION

Read the discussion which follows carefully before attempting to complete the exercise. Also read the appropriate chapter in your textbook.

DISCUSSION

The base pairs in a DNA molecule are grouped in threes to form triplets or mRNA **codons**. Each codon is specific for a particular amino acid. The sequence of bases on the DNA molecule determines the particular protein to be synthesized. At the end of a protein codon sequence, there are nonsense triplets or codons which function to end the message, very much like punctuation marks end a sentence.

The DNA molecule does not synthesize protein directly because protein synthesis occurs in the cytoplasm on ribosomes. The amino acids necessary to make the protein molecule are also in the cytoplasm. The DNA molecule is confined to the nucleus. Intermediate compounds are necessary to facilitate protein synthesis since they must be used to transmit the DNA code to the ribosomes as well as select free amino acids from the cytoplasm to appear in the protein.

RNA (ribonucleic acid) serves as an intermediate compound for the DNA code. RNA is a single stranded polymer of nucleotides with either some or all of its bases exposed to cytoplasm. It consists of nucleotides of ribose, phosphate, and the nitrogen–containing bases adenine, guanine, cytosine, and uracil. The major differences between DNA and RNA are the types of sugar present (deoxyribose vs ribose) and the substitution of uracil for thymine in RNA. RNA is made in the nucleus of the cell. The RNA molecule migrates from the nucleus to the cytoplasm.

There are several types of RNA. Nucleolar RNA is a temporary storage type of new RNA prior to its migration into the cytoplasm. Ribosomal RNA makes up the ribosome. Messenger RNA (mRNA) reads the genetic code for a particular protein off the DNA molecule and transfers this information to the ribosome. Transfer RNA (tRNA) is responsible for the physical selection of amino acids from the cytoplasm and their transport to the ribosomes.

Messenger RNA or mRNA reads the DNA code by pairing up opposite base pairs on a section of the DNA molecule that has split apart from its other strand. For example, the DNA base adenine will pair with the RNA base uracil. The DNA base cytosine will pair with the RNA base guanine and so on. Once the RNA bases are lined up according to the DNA sequence of codons for a particular protein, the mRNA strand leaves the nucleus and goes to the cytoplasm where it attaches to a ribosome. Once attached to the ribosome, it begins to attract specific tRNA units. The function of the mRNA molecule may be compared to that of a secretary transcribing dictation from the boss DNA.

Transfer RNA or tRNA in the cytoplasm is shaped like a hairpin. At the open end it has an attachment site of bases (adenine–cytosine) for amino acids. At the closed end, it has an anti– codon triplet for a specific codon from the DNA molecule. Each anti–codon consists of the opposing bases for a codon of bases. The anti–codon triplet is specific for one of the amino acids, insuring correct positioning of transported amino acids.

As the mRNA molecule moves through the ribosomes, tRNA reads off the code, depositing its specific amino acid in the correct sequence to form a particular protein. The mRNA is then broken down into free nucleotides which can be used to make more RNA strands. The tRNA is released to go back into the cytoplasm to pick up amino acids for future protein synthesis.

LABORATORY EXERCISE PROCEDURE

Carefully follow the directions for each of the following experiments, recording observations, conclusions and answers on the appropriate section of the report sheet. When the experiments are complete submit the completed report sheet to the instructor for grading. Also read the appropriate chapter(s) in your textbook.

LABORATORY EXERCISES

1. Given a DNA strand of codons, (Figure 1) list the mRNA transcription of the code by listing the sequence of bases in mRNA complimentary to the DNA.

   _____

   _____

2. Determine the order of amino acids in the protein by matching up the tRNA anticodon with its corresponding mRNA codon. See Figure 2.

_____

_____

3. Match the names of the amino acids (Figure 3) with their numbers from Figure 2.

_____

_____

4. Now show the shape of the protein. The protein is oxytocin, a hormone that acts on the muscles of the uterus, causing them to contract. This protein is actually not a straight ine of amino acids. Cysteine in the number 4 position bonds to the cysteine in the number 9 position by a disulfide bond.

5. Transfer the answers to Section I of the Report Sheet.

6. Complete Section II on the Report Sheet.

DNA-strand	mRNA	tRNA Codons from Figure 2	Amino Acids (number and name) From Figure 3
C C T	G G A	Answer to IA first blank	C C U	4 Glycine	Answer to IB, C first blank
G A A			
G G T			
A C G			
T T A			
G T T			
G A C			
A T G			
A C G			

FIGURE 1

| UUA | ACG | GUU | CCU |
| 1 | 2 | 3 | 4 |

| GAC | GAA | GGU | AUG |
| 5 | 6 | 7 | 8 |

FIGURE 2
tRNA Codons

1. Asparagine
2. Cysteine
3. Glutamine
4. Glycine
5. Isoleucine
6. Leucine
7. Ptoline
8. Tyrosine

FIGURE 3
Amino Acids

RNA AND PROTEIN SYNTHESIS

Report Sheet 1

Name _____

Student ID # _____

Campus _____

Date _____

I. Synthesizing a Protein

   A. List the sequence of bases in mRNA complimentary to the DNA in Figure 1.

   _____   _____

   _____   _____

   _____   _____

   _____   _____

   B. List in order the amino acids in the protein being synthesized by matching tRNA anti–codons from Figure 2 with the corresponding mRNA codon Figure 1. (These answers will be numbers.)

   _____   _____

   _____   _____

   _____   _____

   _____   _____

   _____

   C. List the names of the amino acids (in order) from Figure 3 with their numbers from answer B.

   _____   _____

   _____   _____

   _____   _____

   _____   _____

   _____

RNA and Protein Synthesis

RNA and Protein Synthesis　　　　Student ID # _____
Report Sheet 2

　　D.　Sketch the protein synthesized

II.　Discussion Questions

　　A.　List the bases in RNA.

　　　　1. _____　　　　3. _____

　　　　2. _____　　　　4. _____

　　B.　List the four types of RNA.

　　　　1. _____　　3. _____

　　　　2. _____　　4. _____

　　C.　Answer these questions about the synthesis of a protein.

　　　　1.　What is a codon? _____
　　　　　　_____

　　　　2.　What is an anti–codon? _____
　　　　　　_____

　　　　3.　What tRNA bases can be attached to the mRNA codon, UGC?
　　　　　　_____

82　Exercise 9

RNA and Protein Synthesis　　　　　　　Student ID # _____
Report Sheet 3

4. Where do the amino acids available in the cytoplasm of a cell come from?

   _____

D. Briefly describe how the amino acids are combined to form a protein. _____

   _____
   _____
   _____
   _____
   _____

E. Provided below is an open strand of DNA located in the nucleus. mRNA is being synthesized. Complete the mRNA by providing the proper nitrogen bases.

   _____ DNA
   TAC　　GGG　　TCC　　ACA　　AAA　ATA

   AUG _____ mRNA

F. The mRNA illustrated below has attached itself to ribosomes in the cytoplasm and the process of protein synthesis is occurring. Complete the anti–codons for the tRNAs.

   　　　　　　　　　　　　　　　　　　　　　　　　　　amino
   　1　　　2　　　3　　　4　　　5　　　6　　　acids
   GUA　　　　　　　　　　　　　　　　　　　　　　　　tRNA

   CAU　　GUA　　AAU　　UGA　　GGG　　CUU　　mRNA

RNA and Protein Synthesis Student ID # _____
Report Sheet 4

G. From the list provided below of codons and the amino acids they code for, list the six amino acids in the protein chain above starting at the left of the tRNA chain.

    AAC—Asparagine   GAA—Glutamic Acid
    ACU—Threonine   GCC—Alanine
    ACG—Serine   GUA—Valine
    CAU—Histidine   UAU—Cysteine
    CAG—Glutamine   UGU—Cysteine
    CCC—Proline   UUA—Leucine

Amino Acids _____, _____, _____,

_____, _____, _____,

The list of 6 amino acids you have just made illustrates the correct sequence of the first 6 amino acids in one of the polypeptide strands of normal hemoglobin.

H. The textbook discusses the types of mutations which can occur in chromosomes and the genes they carry—point mutations, chromosomal changes, and crossing over. Sickle–cell anemia results when the substitution of valine for glutamic acid in position six on one of the polypeptide chains of hemoglobin occurs. List the first six amino acids and their codons as found in the hemoglobin of a sickle cell anemia patient.

              Amino Acid   Codon

1. _____   _____

2. _____   _____

3. _____   _____

4. _____   _____

5. _____   _____

6. _____   _____

What type of mutation is sickle–cell anemia? _____

10 EVOLUTION

LESSON OBJECTIVES

Upon completion of this laboratory the student will be able to:

1. Determine how natural selection can affect evolution.

2. Determine the effect of predators on the evolution of the prey organism.

3. Determine the role of the environment in natural selection.

4. Describe how different organisms can survive in different environments.

5. Understand and use the Hardy–Weinberg formula.

6. Figure gene frequency from a population sample.

7. Use the Hardy–Weinberg formula to determine genotype frequencies.

MATERIALS NEEDED

1. Laboratory Manual
2. Textbook
3. Newspaper sheets (the artificial environments)
4. Circles or squares cut from both solid white and black paper
5. Forceps*
6. Dishes containing 60 brown bean and 40 white beans (2 packages of dried beans of different colors will also work)

   *This item is found in your lab kit.

PREPARATION

Read the discussion which follows as well as the appropriate chapter(s) in your textbook.

DISCUSSION

Evolution

Evolution means change. The study of evolution involves studying the process whereby species change either physically or physiologically through time and thus become adapted to their environment. The mechanism of evolution is natural selection which is the survival of the fittest. Specifically, their are different selecting agents in the environment, which function to either remove organisms directly from the environment by death and/or illness or to produce a situation that is not favorable for maximum breeding. Survival of the fittest simply means that those organisms which have adapted to the environment and have survived predation, parasitism, disease, competition and the changes in the environment can reproduce and leave more offspring thereby ensuring the continuation of the species. Because these individuals can survive and can thus leave more offspring, in the following generations there will be more individuals that carry the genes for those characteristics that provided the survival advantage.

Population Genetics

A population is a group of individuals surviving together in the same place at the same time. "Surviving" implies that the individuals are interbreeding and forming future generations. This exercise is concerned with the genetic relationship of a population from generation to generation.

Consider a population in which 60% of the genes are "A" and 40% are "a".

These alleles can exist in three different combinations: AA, Aa, or aa. The relative frequency of each of these genotypes is dependent upon the relative abundance of each of the alleles. Note the following:

p = the frequency of the allele A = 60% = 0.6
q = the frequency of the allele a = 40% = 0.4
$p + q$ = 60% + 40% = 0.6 + 0.4 = 1.0

If p is the frequency of A, what is the frequency of AA? Multiplying $p \times p$ or p^2 gives us the answer. What are the frequencies of the other genotypes?

By substituting in the values of p and q into the expanded binomial, we can find the actual frequencies of each genotype in the population.

$p^2 + 2pq + q^2 = 1$
$(0.6)^2 + 2(0.6 + 0.4) + (0.4)^2 = 1$
$0.36 + 0.48 + 0.16 = 1$

What are the frequencies of the alleles that this generation could theoretically produce?

AA = 0.36, that is, all A alleles in the gametes
aa = 0.16, that is, all a alleles in the gametes
Aa = 0.48, that is, 1/2 A alleles and 1/2 a alleles in the gametes

Therefore, the frequencies of the alleles in the gametes are:

A = 0.36 + 1/2(0.48) = 0.36 + 0.24 = 0.60
a = 0.16 = 1/2(0.48) = 0.16 + 0.24 = 0.40

These are the same frequencies that we started out with in the first generation. This situation (called a Hardy-Weinberg equilibrium) will always reach equilibrium in one generation and result in the constancy of the relative allelic frequencies AS LONG AS THE FOLLOWING FIVE PARAMETERS ARE MAINTAINED:

1. A large population size
2. Random mating

3. No selection occurs
4. No differential migration occurs; that is, the relative movement of individuals carrying each of the alleles into and out of the population is equal
5. No mutation occurs

By changing any one of these 5 parameters, a change in the gene frequencies in the next generation will occur: this is evolution.

LABORATORY EXERCISE PROCEDURE

Carefully follow the direction for each of the following exercises, recording observations, conclusions and answers on the appropriate sections of the report sheets. When the exercise is complete, submit the completed report sheets to the instructor for grading.

LABORATORY EXERCISES

I. Demonstration of Natural Selection of Prey and Predation

   A. Obtain an artificial environment (a sheet of newspaper) and ten circles or squares cut rom newsprint. The circles or squares will represent the polymorphic (more forms) prey species.

   B. Mix the circles (or squares) and scatter them across the environment.

   C. As the predator, capture (pick up the circles with one hand) the prey and consume (put the circles in a container off to one side) the prey organisms, *one at a time,* for one minute.

   D. Reproduce the surviving prey organisms by adding to the environment two circles of the same color as the surviving circles.

   E. Repeat the predation and reproduction process for 4 generations and enter the number of survivors after each generation in the appropriate table in the lab report.

   F. Repeat the process using forceps as a different predator technique. Record the results.

II. Effect of Mutation

   A. Change the prey by using circles or squares cut from solid white and solid black paper.

   B. Select the type of predation technique that was most successful and repeat the predation and reproduction process for 4 generations of the mutated prey. Record your results.

III. Demonstration of Hardy–Weinberg

  A. Count out into a dish, 60 brown beans and 40 white beans. This is the initial population or gene pool with which we will work. Let p represent the brown beans (alleles) and q represent the white. The brown is a dominant allele; the white, a recessive allele.

  The allele frequency is: p = 0.60 and q = 0.4

  B. Random breeding: no selection or mutation

  1. Stir the beans and without looking pick out two beans. This is an individual. If it consists of two brown beans, record it as a homozygous dominant (p^2 or AA). If both are white, record it as a homozygous recessive (q^2 or aa). If one is white and the other brown, record it as the heterozygote (pq or Aa).

  2. Put the two beans back into the dish (so as not to change the total gene pool size) and repeat the process, recording the results again.

  3. When you have recorded 50 individuals, add up the p^2 and q^2 and pq columns. This gives the number of individuals with each of the possible genotypes. From this determine the total number of p alleles and q alleles in these 50 individuals. How close is the allele frequency to the original p = 0.6 and q = 0.4 with which you started.

While one would expect the frequency to remain the same in a larger population, in a smallpopulation it may vary. The extent that the frequency varies from the original 60:40 ratio, is due to genetic drift. The gene frequency of the population has changed during this one generation without any selection or factors other than chance being involved.

IV. 50 % Selection Against the Dominant Phenotype

  A. Prepare a 0.6 brown: 0.4 white gene pool with the beans as you did above.

  B. Again record 50 individuals at random but this time do not count every other genotype which is either AA or Aa (p^2 or pq). You will actually choose more than 50 pairs of beans as you want to record 50 pairs. This is as if 50% of the dominant phenotypes did not survive.

  C. Count up the total number of each genotype recorded and determine the gene frequency of this generation. Can you see any effect of the selection? Record your answer on the answer sheet, Section IV, A.

  D. Now adjust your gene pool (dish of beans) to the results you got; that is, if you got a frequency of p = 55 and q = 45 on the first generation, then your dish of beans should now be made to contain 55 brown and 45 white beans. You are now ready to select another generation.

  E. Using the new gene pool, select and again record 50 pairs of genes. Again do not record every other pair which contains a brown bean (dominant phenotype), but do record 50 pairs.

88 Exercise 10

F. Again add up and determine the allele frequency. Is there further change from the initial frequency? Record your answer on the lab report, Section IV. B.

G. Record your results and again compare the individual groups of 50 with the larger population. Using the total figures, is the change during the second selection as great as, greater than, or the same as that during the first generation of selection. Record your answer on the lab report, Section IV, C. If you were to continue several more generations of selection done in the same way, how many generations would it take to reach a zero frequency of one of the alleles and 100% frequency of the other? Record your answer on the lab report, Section IV, D.

H. While 50% selection is very high, remember this is a very small population. If 50% selection occurred in a very large population, how might you expect the relative effects of drift and selection to differ? Record your answer on the lab report, Section IV, E.

I. How do the effects of genetic drift and selection compare? Record your answer on the lab report Section IV, F. By changing any one of these 5 parameters, a change in the gene frequencies in the next generation will occur: this is evolution.

EVOLUTION

Report Sheet 1

Name _____

Student ID # _____

Campus _____

Date _____

Part I.

I. Demonstration of Natural Selection of Prey and Predator

Hand used by predator:

Number of Survivors

	Trial 1	Trial 2	Trial 3	Trial 4
1st generation				
2nd generation				
3rd generation				
4th generation				

Forceps used by predator:

Number of Survivors

	Trial 1	Trial 2	Trial 3	Trial 4
1st generation				
2nd generation				
3rd generation				
4th generation				

II. Effect of Mutations

Number of Survivors

	White 1	White 2	White 3	White 4
1st generation				
2nd generation				
3rd generation				
4th generation				

Evolution
Report Sheet 2

Student ID # _____

Answer the following questions:

A. What was the role of color in adaptation of the prey species? _____

_____

How did it correlate with the environment? _____

_____

_____

B. Did any color prey ever become extinct? _____ Why or why not? _____

_____

_____

C. What predator technique was most successful? _____ Why? _____

_____

_____

D. What is meant by co–evolution of predator and prey?

_____

_____

E. What effect did mutations have on the prey species?

_____

_____

F. Can you hypothesize what effect a mutation might have on the predator in nature?

_____

_____

_____

Evolution
Report Sheet 3

Student ID # _____

III. Demonstration of Hardy–Weinberg

    A. Was the drift in the same direction and of the same degree each time you performed the experiment?

    _____
    _____

    B. How much drift does the total population show?

    _____
    _____

    C. As the population got larger and larger, how would you expect the importance of the drift factor to change?

    _____
    _____

IV. 50% Selection Against the Dominant Phenotype

    A. Can you see any effect of the selection?

    _____

    Explain _____
    _____

    B. Is there further change from the initial frequency? _____

    Explain _____
    _____

    C. Is the change during the second selection as great as, greater than, or the same as that during the first generation of selection?

    _____

    Explain _____
    _____

Evolution
Report Sheet 4

Student ID # _____

D. If you were to continue several more generations of selection done in the same way, how many generations would it take to reach a zero frequency of one of the alleles and 100% frequency of the other?

_____

_____

_____

E. If 50% selection occurred in a very large population, how might you expect the relative effects of drift and selection to differ?

_____

_____

_____

F. How do the effects of genetic drift and selection compare? _____

_____

_____

11 CLASSIFICATION

LESSON OBJECTIVES

Upon completion of this laboratory exercise the student should be able to:

1. Define or identify the following: taxonomy, Linnaeus, binomial system of nomenclature, species.

2. Write a scientific name correctly.

3. List in order the seven categories in the classification system.

4. List the five kingdoms of living organisms and give an example of each.

5. Devise a classification system for a group of objects.

6. Identify organisms using an identification key.

MATERIALS NEEDED

1. Laboratory manual
2. Pen or pencil

PREPARATION

Read the discussion which follows carefully and the appropriate chapter(s) in your textbook before attempting to complete the exercise.

DISCUSSION

There are around two million species of organisms living on this planet. At least 1 1/2 million of these organisms have been named already and others continue to be described each year. Classification seems necessary due to the large number of species of organisms and the necessity of communicating with others about the different species.

The ancient Greeks, notably Aristotle, laid the foundations for **taxonomy**, the study of the classification of organisms. Our modern system of classification was developed by **Carl Linnaeus**, a Swedish botanist. The Linnaean system is known as the **binomial system of nomenclature** because it assigns two names to each organism. These two names are the **genus** and **species** names.

Linnaeus's system indicates that one species can be related to other species. A **species** is defined as a group of naturally interbreeding organisms, a genetically distinctive population isolated reproductively from all others. A group of related species can form a higher category called a **genus**. Several genera could be grouped into a higher category and so on up the scale. The value of this type of system is that given the generic and specific name for an organism, its relationship to other species within that genus can be determined and its relationship to categories higher than the genus can also be determined.

Scientific names generally are composed from Greek or Latin. Latin is a very descriptive language and the words tend to define themselves. Also, Latin is unchanging. The scientific names may be a Latinization of the geographic area where the organism is found, the color or size of the organism, or any outstanding characteristic of the organism. The generic name is generally a noun and the specific name is generally an adjective describing the noun. By convention, the genus name is always capitalized, while the species name is not. Both genus and species names are underlined or italicized.

There are seven classification categories: kingdom, phylum, class, order, family, genus, and species. Each organism is grouped into these categories based on the characteristics it shares with members of the same category. Each organism is grouped into one of five kingdoms. Organisms without a nuclear membrane belong to kingdom **Monera**, single-celled organisms with nuclear membranes to kingdom **Protista**, fungus organisms to kingdom **Fungi**, multicellular plants to **Plantae** and multicellular animals to kingdom **Animalia**.

It is not always easy to identify organisms. Sometimes the differences between two species of organisms are very slight. Biologists use identification keys (also called couplet keys) to help them identify organisms. Usually an identification key is made up so that as you read it, you have two choices for each characteristic listed. There could be a key to evergreen tree genera where you would have to decide whether a tree has needles wrapped in bundles or not wrapped in bundles. Each choice you make in the identification key leads you to further choices more specifically identifying the organism. Eventually the organism is "keyed out" or named.

LABORATORY EXERCISE PROCEDURE

Carefully complete each exercise as indicated and record the required information on the report sheets. When complete, submit the report sheets to the instructor for grading.

FIGURE 1

Classification 97

Classification Key to Certain Mythological Creatures

Choice	Characteristic	Next Move or Identification
1a	The creature has the body of a horse	2
1b	The creature does not have the body of a horse	4
2a	The creature has wings	Pegasus
2b	The creature does not have wings	3
3a	The creature has the head of a man	Centaur
3b	The creature has a horn	Unicorn
4a	The creature has the tail of a fish	5
4b	The creature does not have the tail of a fish	7
5a	The creature has four legs	Sea Serpent
5b	The creature has two legs or two arms	6
6a	The creature has the head of a horse	ooved–Fish
6b	The creature has the head of a man	Man–Fish
7a	The creature has wings	8
7b	The creature does not have wings	Scaled–Lion
8a	The creature has the head of a man	Winged–Sphinx
8b	The creature has the head of a rooster	Winged–Chicken

Exercise 11

CLASSIFICATION

Report Sheet 1

Name _____

Student ID # _____

Campus _____

Date _____

I. A Classification Key for Mythological Creatures

   Objective: Set up a classification system using Kingdom, Phylum, and Class

   Materials: Pictures of some mythological creatures

   Procedure:

   A. Place the objects into two "kingdoms" based on the characteristics they share in common. Kingdom 1 should contains objects 1, 2, 4, 5, and 8. Kingdom 2 will contain objects 3, 6, 7, and 9, using the presence or absence of body scales as a basis of separation.

   What could Kingdom 1 be named? _____

   What could Kingdom 2 be named? _____

   B. Now divide Kingdom 1 into 2 Phyla. Phylum 1 should have 4, 5, and 8 in it. Phylum 2 will have 1 and 2.

   What characteristic is shared by all members of Phylum 1? _____

   _____

   Phylum 2? _____

   _____

   C. Further divide each Phylum into 2 classes. What objects were placed in class 1 of :

   Phylum 1? _____

   Class 2? _____

   Why? _____

   D. Now divide Kingdom 2 in Phyla. Phylum 1 will have 6, 7, and 0. Phylum 2 has only #3. What characteristic does each member of Phylum 1 share?

   _____

   _____

Classification
Report Sheet 2

Student ID # _____

    E. Now divide the members of Phylum 1 into two classes. Class 1 contains_____

        Class 2 contains _____ Why? _____

        _____

II. Classification Using a Key

    Objectives: Key out some organisms using a couplet type of identification key.

    Materials: Classification key and pictures to be classified

    Procedure:

    A. Carefully study the pictures which are to be "keyed out."

    B. Beginning with 1a on the classification key, read the statement and determine if it describes the picture. If it does, then go to the *next move* or *identification* to the right of the characteristic and follow its instructions. If 1a does not describe the picture, then skip to 1b. Follow the same procedure until the identification is complete. The first picture will be keyed out as an example.

        Look at picture 1. Now look at the key. 1a fits the picture; therefore the next move is to 2, skipping 1b. 2a does not fit but 2b does. Go to 3. 3a fits the picture. Picture 1 is identified as a Centaur.

    C. List the names of the picture "keyed out."

        1. _____   6. _____

        2. _____   7. _____

        3. _____   8. _____

        4. _____   9. _____

        5. _____

III. Answer the following:

    A. Give a definition for each of the following:

        1. Binomial system of nomenclature _____

        _____

Classification
Report Sheet 3

Student ID # _____

2. Linnaeus _____
   _____

3. Taxonomy _____
   _____

4. Species _____
   _____

B. Write the following scientific name correctly:

   genus name: Rhinoceros
   species name: Bicornis

   _____

C. List the seven categories of classification in order starting with the most general group.

   1. _____   5. _____
   2. _____   6. _____
   3. _____   7. _____
   4. _____

D. List the five kingdoms of living organisms and give an example of each.

   1. _____
   2. _____
   3. _____
   4. _____
   5. _____

Classification Student ID # _____
Report Sheet 4

E. Questions using the key from Report Sheet 1.

1. Upon what is this entire classification key based? _____
   _____
   _____

2. Why is it important to separate organisms into groups? _____
   _____
   _____

12 VIRUSES, BACTERIA, AND PROTISTANS

LESSON OBJECTIVES

Upon completion of this laboratory exercise the student will be able to:

1. Describe the structure of a virus.
2. List several diseases caused by viruses.
3. Discuss the replication of viruses.
4. Identify the three basic types of bacteria.
5. List several ways in which bacteria are of benefit to living organisms.
6. List 2 ways in which bacteria are harmful to living organisms.
7. Explain why the protists are not classified as plants or animals.
8. Give examples of both plant–like and animal–like protists.

MATERIALS NEEDED

1. Laboratory manual
2. Textbook
3. *Photo Atlas for Biology*

PREPARATION

Read the discussion which follows carefully before attempting to complete the exercise. Also read the appropriate chapter in your textbook.

DISCUSSION

Viruses

Viruses are a nucleic acid particle encased in a protein coat. They can only function inside a living cell so are not considered to be living. They are host–specific and can only infect a cell that has the proper receptor sites to which the virus can attach. Viruses are able, inside a host cell, to take control of the host's metabolic pathways and direct it to carry out the work of making new viruses. The new viruses are then released from the cell to invade new host cells. Most viruses are identified by their activities in host cells. Because of their very small size, most viruses require an electron microscope to be seen.

Bacteria

Bacteria are small, single–celled organisms with cell walls containing organic molecules not found in other kinds of organisms. These organisms, classified in Kingdom Monera, are called prokaryotes because they have no nucleus and the genome is a single strand of DNA. Bacteria do not undergo mitosis or meiosis.

Many forms of bacteria are beneficial to man. Some bacteria are saprophytes, organisms that obtain energy by the decomposition of dead organic material; others are parasites that obtain energy and nutrients from living hosts. Other bacteria are mutualistic or commensalistic with host organisms.

Protists

Members of Kingdom Protista are also one–celled organisms. The protists are eukaryotes. Eukaryotes have a distinct nucleus and other cell organelles such as mitochondria, endoplasmic reticulum and chloroplasts. All eukaryote species can undergo mitosis; some can also undergo meiosis. Many eukaryotes contain chlorophyll in chloroplasts and are autotrophic. Other eukaryotes require organic molecules as a source of energy and are heterotrophic.

The protists are usually divided into three groups: the algae (autotrophic unicellular organisms), the protozoa (heterotrophic unicellular organisms), and fungus–like protists. The algae are protists with a cellulose cell wall that contain chlorophyll and carry on photosynthesis. The protozoans are classified according to their locomotion. They have no cell walls and no chloroplasts. The fungus–like protists have a motile amoeboid reproductive stage and are classified as slime molds or water molds.

LABORATORY EXERCISE PROCEDURE

Carefully read the information in the appropriate chapter(s) in your textbook before attempting the following exercises. Complete each exercise as indicated on the report sheets. When complete, submit the report sheets to the instructor for grading.

LABORATORY EXERCISES

I. The Virus—*Photo Atlas for Biology*, fig. 5c, pg. 5

   A. Look at fig. 5c and determine how the virus is attached to the bacterium. Answer the question on the lab report relating to the attachment.

   B. After observing fig. 5c, sketch a typical bacteriophage and label the head and tail. Space is provided in the lab report for the sketch. The sketch must resemble what you see in the electron micrograph, not the drawing.

   C. Explain why a cloudy culture containing bacteria clears quickly when bacteriophages are added.

   D. List 6 animal diseases caused by viruses.

II. Bacteria—*Photo Atlas for Biology,* figs. 17c & 17d, pg. 17.

   A. After observing figures 17c and 17d, identify one of the shapes that bacteria take.

   B. List five ways bacteria are of benefit.

III. The Kingdom Protista—Refer to *Photo Atlas for Biology*

   A. Draw and label an amoeba (fig. 22a, pg. 22). Include cell membrane, nucleus, vacuole, nd pseudopod in your labels. Give a function for each organelle.

   B. Draw and label a paramecium (fig. 22e, pg. 22). Include cell membrane, macronucleus, micronucleus, oral groove, and cilia in your labels. Give a function for each organelle.

   C. Draw and label an Euglena (fig. 18d, pg. 18). Include chloroplasts, contractile vacuole, flagellum, cell membrane, nucleus and eye spot in your labels. Give a function for each organelle.

   D. Draw and label one cell of a Spirogyra (fig. 36a–e, pg. 36). Include cell wall, cell membrane, chloroplasts, and cytoplasm in your labels. Give a function of each organelle.

IV. Answer all questions in the lab exercise.

VIRUSES, BACTERIA, AND PROTISTANS

Report Sheet 1

Name _____

Student ID # _____

Campus _____

Date _____

Note: All sketches **must** reflect the material as it appears in the photos. Copying drawings for any other source is not acceptable. Credit will not be given for drawings from other sources or for drawings on the wrong page or place on the page.

I. The Viruses

    A. Which part of the virus is attached to the bacterium? _____

    _____

    B. Sketch a bacteriophage:

    C. Explain why a cloudy culture containing bacteria clears quickly when bacteriophages are added.

_____

_____

_____

_____

_____

_____

Exercise 12

Viruses, Bacteria, and Protistans   Student ID # _____
Report Sheet 2

    D. List six animal diseases caused by viruses:

        1. _____   4. _____

        2. _____   5. _____

        3. _____   6. _____

II. Bacteria

    A. Identify two bacterial shapes: _____

    B. List five ways bacteria are of benefit:

        1. _____   4. _____

        2. _____   5. _____

        3. _____

III. The Kingdom Protista

    A. Draw and label an amoeba:

Viruses, Bacteria, and Protistans   Student ID # _____
Report Sheet 3

    B.   Draw and label a paramecium

    C.   Draw and label an Euglena

    D.   Draw and label one cell of Spirogyra

Viruses, Bacteria, and Protistans   Student ID # _____
Report Sheet 4

IV. Analysis and Conclusions

　　A. Based on what you have learned about viruses, why are viruses not considered to be alive?

　　　　_____

　　　　_____

　　　　_____

　　B. Where are viruses produced? _____

　　　　_____

　　C. List the three types of bacteria. You will need to refer to your textbook for help.

　　　　1. _____ 2. _____ 3. _____

　　D. Why are some bacteria, such as the pneumonia bacteria, so resistant to the body's natural defenses against disease?

　　　　_____

　　　　_____

　　　　_____

　　E. Where are bacteria found? _____

　　F. Compare and contrast the following methods of locomotion by protozoa.

　　　　1. Pseudopod _____

　　　　　　_____

　　　　2. Cilia _____

　　　　　　_____

　　　　3. Flagella _____

　　　　　　_____

Viruses, Bacteria, and Protistans   **109**

Viruses, Bacteria, and Protistans   Student ID # _____
Report Sheet 5

G. Why is an Euglena considered to be both plant–like and animal–like? _____
   _____
   _____

H. Describe the life cycle of *Plasmodium vivax*. _____
   _____
   _____
   _____

I. Define symbiosis. _____
   _____
   _____

J. Give an example of symbiosis. _____
   _____
   _____

K. Spirogyra has two methods of reproduction. Briefly describe them.

   Method 1 _____
   _____
   _____

   Method 2 _____
   _____
   _____

Viruses, Bacteria, and Protistans   Student ID # _____
Report Sheet 6

L. What are the "red tides" and why are they a problem? _____

_____

_____

_____

M. What do diatoms produce that the other protists do not? _____

_____

_____

13   FUNGI AND PLANTS

LESSON OBJECTIVES

Upon completion of this laboratory exercise the student will be able to:

1. Define the term fungi.

2. Describe the process of budding.

3. Describe the structure of a mushroom.

4. Define alternation of generations and explain the life cycle of a moss plant.

5. Define the terms gametophyte, sporophyte, antherid and archegone.

6. Compare the life cycle of a fern with the life cycle of a moss, noting the significance of the sporophyte and gametophyte generations.

MATERIALS NEEDED

1. Laboratory Manual
2. Textbook
3. *Photo Atlas for Biology*

PREPARATION

Read the discussion which follows as well as the appropriate chapter(s) in your textbook.

DISCUSSION

Fungi

Members of Kingdom Mycetae, the true fungi, are nonphotosynthetic, eukaryotic organisms. The fungi were originally classified in the Plant Kingdom, but are now in their own kingdom because they do not contain chlorophyll and do not photosynthesize. They absorb food that they predigest with enzymes. Because of their mode of food–getting, the fungi play a significant role in the biosphere as decomposers and recyclers. The fungi do have a cell wall but its composition is different from that of plants.

The first division of fungi are the members of Zygomycota, referred to as **zygomycetes** because they produce sexual spores called **zygospores**. A common zygomycete is the black bread mold. It attaches to bread with a mycelium, a tangled mass of threads called hyphae. Certain hyphae develop a sporangium which contains the asexual spores of this fungus.

Zygospores are only produced when two different mating types grow into contact with each other.

The second division of fungi is called the Ascomycota. These fungi are sometimes called sac fungi because their sexual spores are produced in sacs called **asci**. The yeast are examples of Division Ascomycota. The **ascomycetes** also include mildews, morels and truffles. Asexually they reproduce when conidia (spores) pinch off from conidiophores.

The Basidiomycota make up the third division of fungi and include mushrooms, bracket fungi, puffballs, rusts and smuts. Asexual reproduction is uncommon in **basidiomycetes**. Basidiospores are produced from a basidium during sexual reproduction.

The fourth division of fungi is the Deuteromycota or imperfect fungi. These fungi have no identified sexual stage in their life cycle. Most reproduce only by means of conidia (spores).

Plants

Complex photosynthetic organisms are placed in the **Kingdom Plantae**. Members of Kingdom Plantae include the mosses, liverworts, ferns, angiosperms and gymnosperms. Some biologists also include green, red and brown algae in this kingdom; others, place all algae in Kingdom Protista. One of the most important adaptations plants have in order to survive on land is a waxy coating, the **cuticle**, over their aerial parts. The cuticle helps prevent evaporation from drying out plant tissue.

Plants have multicellular sex organs or **gametangia**. The female organ, the **archegonium**, produces a single egg. Sperm are produced in the male sex organ, the **antheridium**. The embryo is protected within the female gametangium. Algae lack such sexual organs or protection for their embryos.

Plants have a clearly defined **alternation of generations**. The **gametophyte generation** gives rise to gametes by mitosis. The **sporophyte generation** gives rise to spores by meiosis. The gametophytic plant produces antheridia and archegonia. Sperm get to the archegonia in a variety of ways, and one sperm fertilizes one egg resulting in a **zygote**. The zygote is the first cell in the sporophyte generation. It develops and eventually matures into the sporophyte plant. The sporophyte plant has special cells capable of division by meiosis. These cells, called the **spore mother cells**, undergo meiotic division and form spores. The spores represent the first stage in the gametophyte generation.

The more primitive plants are the **bryophytes**, the only nonvascular plants. Vascular plants have specialized tissues for the transport of water, food and other nutrients. Since the bryophytes lack a vascular system, they are severely limited in size. Bryophytes include the mosses, liverworts, and hornwarts. They exhibit alternation of generations, as do higher plants. The gametophyte generation is considered the dominant generation. The sporophyte generation is at all times attached to and dependent on the gametophyte plant.

The **ferns** show the first presence of specialized vascular tissue. Most ferns have true roots, stems and leaves. Their life cycle is also an alternation of generations; however, the sporophyte generation is now the dominant generation. Water is still needed for fertilization, however. The gametophyte plant of ferns bears no resemblance to the sporophyte plant.

The primary means of reproduction and dispersal for the most successful plants is by seeds, which develop from the female gametophyte and tissues associated with it. The seed plants are the **gymnosperms** and **angiosperms**. The gymnosperms produce seeds that are totally exposed or borne on the scales of cones. Pine, spruce, fir and other conifers are examples of gymnosperms. Angiosperms, or flowering plants, produce their seeds within a fruit. Both gymnosperms and angiosperms possess vascular tissue. Both have alternation of generations with the gametophyte significantly reduced in size and totally dependent on the sporophyte generation. They are both heterosporous, producing microspores which eventually produce male gametophytes and macrospores which produce female gametophytes and eggs.

LABORATORY EXERCISE PROCEDURE

Carefully complete each exercise recording the required information on the report sheets. When complete, submit the report sheets to the instructor for grading.

LABORATORY EXERCISES

I. Fungi—*Photo Atlas for Biology*, pp. 24–31

   A. From the yeast fig. 26a, draw a yeast cell undergoing budding on your lab report.

   B. Observe the bread mold in fig. 24b. Sketch a bread mold, labeling the stolen, rhizoids, sporangiophore and sporangium.

   C. Observe the mushroom in figs. 28a & b. Sketch a pie–shaped wedge of the mushroom cap, labeling the gills and spores.

II. Plants

   A. Moss—*Photo Atlas for Biology*, pp. 244–46.

      1. After observing the figures, sketch an archegone, antherid, sporophyte plant and protonema.

      2. Using your textbook to help you, answer the questions on the entire moss life cycle on your lab report.

   B. Fern—*Photo Atlas for Biology*, pp. 51–53

      1. Observe the sporophyte leaf and answer the question on the lab report.

      2. After observing the mature gametophyte, describe its shape. Sketch the gametophyte plant, labeling the root–like structures, the antherid and the archegone.

III. Answer all questions in the lab report.

FUNGI AND PLANTS

Report Sheet 1

Name _____

Student ID # _____

Campus _____

Date _____

I. Fungi

    A. 1. Sketch a budding yeast cell

          a. Why is budding an asexual reproductive process? _____
_____
_____

          b. How do yeast obtain energy? _____
_____

      2. Sketch the Bread Mold

Fungi and Plants
Report Sheet 2

Student ID # _____

Is this an example of the sporophyte generation or gametophyte generation?
_____

3. Sketch a Mushroom

    a. Is this an example of sexual or asexual reproduction? _____

    b. What would these spores be called? _____

II A. Moss

    A1. Sketch the archegone   Sketch the antherid

    Sketch the sporophyte   Sketch the protonema

Fungi and Plants
Report Sheet 3

Student ID # _____

    A2. Moss life cycle: After reviewing the entire life cycle of the moss, given the following structures, list those that are a part of the asexual (sporophyte generation) and those that are sexual (gametophyte generation). Indicate the sporophyte generation by the letters SG and the gametophyte generation by the letters GG.

    _____ female gametophyte   _____ egg

    _____ male gametophyte   _____ mature sporophyte

    _____ male sex organ   _____ capsule

    _____ sperm   _____ spores

    _____ female sex organ   _____ protonema

II. B. Fern

    1a. What are the brown spots on the sporophyte leaf? _____

    _____

    1b. How does the sporophyte reproduce? _____

    _____

    _____

    2. Sketch the gametophyte plant

Fungi and Plants
Report Sheet 4

Student ID # _____

2.a. Which generation is the dominant generation? _____

_____

2.b. Identify each part of the fern life cycle by labeling gametophyte generation (GG) or Sporophyte generation (SG).

_____ mature sporophyte   _____ antherid

_____ spore case   _____ archegone

_____ spores   _____ sperm

_____ mature gametophyte   _____ embryo sporophyte

III. Applications and Conclusions

A. Why are the fungi not classified as plants? _____

_____

_____

_____

B. Mushroom spores are very small and light. How is this an adaptation for reproduction?

_____

_____

_____

C. How do bryophytes differ from ferns? _____

_____

_____

Which generation is the dominant generating in each? _____

_____

14 SURVEY OF THE ANIMAL KINGDOM

LESSON OBJECTIVES

Upon completion of this laboratory exercise the student will be able to:

1. Describe the major evolutionary advances in animal body plans that led to increasingly large and complex animals.

2. Describe distinguishing characteristics of each phylum.

3. Give example organisms of each phylum.

4. Describe the adaptations that have contributed to the great success of the Arthropods.

5. List the two structures present in all Chordates at some time in their life.

MATERIALS NEEDED

1. Laboratory manual
2. *Photo Atlas for Biology*
3. Textbook

PREPARATION

Read the discussion which follows carefully before attempting to complete the exercise. Also read the appropriate chapter or chapters in your textbook.

DISCUSSION

Phylum Porifera

Sponges belong to Phylum Porifera, which means to bear pores, because all members of this group have bodies that contain tiny pores that are basic structures in their functional activity. Sponges are organized at the cellular level. They exhibit radial symmetry or no symmetry. They are attached to a substrate and all are aquatic. The sponge body is a loose aggregation of cells with many pores or ostia, canals and chambers serving for the passage of water. Most of their inner chambers are lined with choanocytes or flagellate collar cells. They have no organs or tissues. They exhibit both sexual and asexual reproduction. Asexual reproduction is by buds or gemmules. There are three classes of sponges based on the type of skeleton they possess. Class Calcispongia have spicules of calcium carbonate. Class Hyalospongiae have six-rayed silicon spicules. Class Demospongiae have silicon and or spongin spicules. Bath sponges are examples of this class.

Phylum Cnidaria

The Cnidarians take their name from the large cavity which serves as an intestine. These organisms demonstrate radial symmetry. They consist of a solid body wall enclosing a gastrovascular cavity. The body wall consists of two well–defined epithelial layers, epidermis and gastrodermis and an intermediate layer, the mesoglea. Members of this phylum exist as attached polyps or free medusa. Their gastrovascular cavity has a single opening and extensible tentacles often encircle the mouth or oral region. Special stinging cells called nematocysts are found on the tentacles. Reproduction is by asexual budding in polyps or sexual reproduction in medusae. The polyp is usually attached to a substrate, while the medusa is a free–swimming jellyfish. Examples include *Obelia* with both a polyp and medusa stage, *Hydra* with only a polyp stage and certain Scyphozoa with only the medusa stage. Classes include Hydrozoa, Scyphozoa and Anthozoa (the sea anemone, corals, and sea pens).

Phylum Platyhelminthes

The flatworms are members of Phylum Platyhelminthes. They are triploblastic with three germ layers. They exhibit bilateral symmetry with definite anterior and posterior ends. They have simple sense organs with eyespots in some. The body of a flatworm is flattened dorsoventrally. The digestive system is incomplete. They have a primitive digestive system with flame cells. Nerve ganglia are present. One class, the Turbellaria, are free–living. Turbellarians reproduce both sexually and asexually. They are hermaphrodic (contain both sexes) but practice cross-fertilization. Planaria have a well developed ability to regenerate lost body parts. Examples of Turbellaria include the planaria.

Class Trematoda include the flukes which are parasitic, requiring one or two hosts for their life history. The human liver fluke is *Opisthorchis sinensis*. Eggs shed from the adult in the bile duct are carried out of the body in feces. There they are ingested by snails, hatch, grow and leave the snail to find a fish host where they burrow under scales to encyst in muscle. When raw or improperly cooked fish containing the cysts is eaten by man, the young flukes enter the bile duct, mature and shed eggs to begin the cycle again. Other common flukes of man are the blood fluke and lung fluke.

Members of Class Cestoda are also parasitic, usually with alternate hosts. Examples include the tapeworm, with its flat body divided into sections called proglottids and a complete lack of digestive system. *Taenia saginata*, the beef tapeworm, lives as an adult in the alimentary tract of man. Ripe proglottids break off in man's intestine, pass out in feces and are ingested by cows. Eggs hatch in the cow's intestine, freeing onchospheres which penetrate into muscles and encyst, developing into "bladder worms." Man eats infected rare beef, the bladder worms are freed in the intestine, develop and attach to the intestine wall.

Phylum Aschelminthes

This phylum, sometimes called Nematoda, has reached the organ–system level of organization. They exhibit a tube–within–a–tube arrangement. They have bilateral symmetry, are unsegmented and triploblastic (three germ layers). Their digestive system is complete with a month, enteron and anus. The most common class is Nematoda.

Nematoda contains the intestinal roundworm, the hookworm, pinworm, filarial worm and guinea worm. The intestinal roundworm, *Ascaris lumbricoides*, is one of the most common parasites found in man. It is pointed at both ends and whitish–yellow in color. The male can be distinguished from the female by its smaller size and by the sharply curved posterior end which bears two genital spicules in the genital pore. In the female the posterior end is straight. The body cavity, in which the visceral organs lie, is called a pseudocoel. The alimentary canal consists of a mouth, pharynx, intestine, rectum and anus. There is a simple excretory system consisting of a lateral excretory canal. Only one host is involved in the life cycle. A female Ascaris lays eggs which are passed with the feces, deposited on the ground, and develop into small worms inside their shells. If taken inside the body through poor sanitation, they will pass to the intestine, hatch into larvae and burrow through the intestinal wall into veins or lymph vessels. From the blood they pass to the lungs, travel up the trachea, cross over to the esophagus where they are swallowed and be in the cycle again. Most damage occurs as the young worms migrate inside the body.

Phylum Mollusca

Members of Phylum Mollusca have a soft body with a ventral muscular foot and fleshy mantle which secretes a shell in most cases. This phylum contains species from the most microscopic to a length of 50 feet, the largest of all invertebrates. Mollusks are found in nearly all places that support life and have a continuous fossil record since Cambrian times. The external body of a mollusk has three typical divisions: an anterior head with mouth, appendages and sense organs; a ventral muscular foot variously modified but generally for locomotion; and a dorsal mantle. The digestive system is complete with digestive glands and liver. A rasping organ, the radula, is usually present. A heart is present with a pericardial space and blood vessels.

The classes of mollusks are based on type of shell, type of foot and shape of shell. The most common classes are Gastropoda, the snails; Pelecypoda, the bivalves such as the oyster and clam; and Cephalopoda, the squids and octopuses.

The fresh water clam has symmetrical right and left valves held together by a thick ligament. Inside the shell is the body, made up of a visceral mass and a muscular foot. The visceral mass is attached to the mantle and is continuous with it. Each lobe of the mantle is attached to a valve and together they form the mantle cavity, which encloses the entire body mass. Posteriorly the mantle is modified to form the dorsal excurrent siphon and the slightly fringed ventral incurrent siphon which regulate the intake and outgo of water. Cilia on the inner surfaces of the siphons and mantle direct the flow of water over two pairs of gills. The digestive system consists of a mouth between two labial palps, an esophagus, a stomach which receives digestive enzymes from the digestive gland (liver), a coiled intestine and a rectum attached to an anus. The circulatory system is an open one with a heart, arteries, sinuses and veins; however, water cannot enter the system. Respiration is carried on by both mantle and gills. Excretion is performed by the C-shaped kidneys just below the pericardium. Coordination is effected by a nervous system of several ganglia scattered throughout the body in pairs. Sexes are separate.

Phylum Annelida

As a group the annelids show definite body systems, a tendency toward centralization of the nervous system with cerebral ganglia (a brain) and nerve cords running the length of the body. The well-developed nephridia have reached a differentiation which involves a removal of waste from the blood as well as from the coelom. The circulatory system is much more complex with a closed system with muscular blood vessels and aortic arches (hearts) for propelling the blood. Annelids are the most highly organized animals capable of regeneration. The introduction of metamerism represents the greatest advancement of this phylum. Metamerism is body segmentation. Circulatory, excretory, nervous, muscular and reproductive organs all show a segmented arrangement. Members of this phylum are hermaphroditic or with separate sexes.

The most common example of Phylum Annelida is the common earthworm, *Lumbricus terrestris*. It is elongated, cylindrical in form and tapering at both ends. Its body is divided into 100–175 metamers (segments). On the anterior side of the first metamere is the mouth overhung by the fleshy prostomium. Straddled over the back like a saddle (metameres 31 to 37) is the clitellium which is important in reproduction. The digestive system consists of the mouth, pharynx, esophagus, crop, gizzard, intestine and anus. The intestine has, on its dorsal wall, an infolded typhlosole which greatly increases the absorptive and digestive surface. The earthworm, although hermaphroditic, does not self–fertilize. Two earthworms line up anterior end to posterior end so that the seminal vesicle openings of one worm are opposite the clitellium of the other worm. Sperm pass out of the vasa deferentia of each worm and travels by seminal canals or grooves on the surface to the openings of the seminal receptacles of the other worm.

Another interesting members of this phylum is the leech, Class Hirudinea. *Hirudo medicinalis* is the medicinal leech used in medical practice for blood letting.

Phylum Arthropoda

Arthropods have the characteristic structure of higher forms; bilateral symmetry, triploblastic, coelomic cavity and organ systems and have reached the peak of invertebrate evolution. They have conspicuous segmentation, especially in the body, muscles and nervous system. Biological contributions of this phylum include cephalization (nervous system control in the head region) and the presence of jointed appendages. Behavior patterns show primitive intelligence and social instincts in some members. Metamorphoresis is common in development.

The arthropods, as a group, are highly successful. They are the most extensive group in the animal kingdom, making up more than three–fourths of all known forms. Some of the structural and physiological patterns which have made this phylum so successful include the presence of chitin. Chitin's structure is adapted for protection of delicate internal organs, for attachment of muscles, for serving as levers and centers of movement, for preventing the loss of and entrance of water and for affording the maximum amount of protection without sacrificing mobility. Segmentation and the presence of jointed appendages has made for greater efficiency and wider capability for adjustment to different habitats. The highly efficient tracheal system of air tubes present in most arthropods delivers the oxygen directly to tissue cells, making possible the high metabolism so characteristic of active insects. No other invertebrate group has the diversity of specialized and delicate sensory organs that arthropods have. These sense organs range from the mosaic eye to organs of touch, smell, hearing, balancing, chemical and others. The jointed appendages have been modified into walking legs, swimming appendages and wings. They lay large numbers of eggs; many pass through metamorphic changes in reaching maturity – larva, papa, and adult stages. This results in less competition for food. Instinctive behavior patterns reach their peak here.

Physically the arthropods have jointed appendages, an exoskeleton of chitin, three divisions of the body; the head, thorax and abdomen, a complete digestive system, an open circulatory system, a respiratory system, excretory system of green glands or Malpighian tubules, a nervous system with a dorsal brain and separate sexes.

The Arthropods are classified into five general classes: Arachnida, the scorpions, spiders, mites and ticks; Crustacea, the crayfish, shrimp barnacles, and crabs; Dipopoda, the millipedes; Chliopoda, the centipedes and Insecta. Insecta is the most successful class of Arthropods.

Phylum Echinodermata

The echinoderms share with the annelids, mollusks and arthropods the distinction of reaching the highest level of organization of invertebrates. This group has radial symmetry as adults but their larval forms are bilaterally symmetrical. The most marked characteristics of echinoderms are (1) the spiny endoskeleton, (2) the water–vascular system, (3) the pedicellariae, (4) the dermal branchiae, (5) the amebocytes and (6) radial symmetry. The water–vascular system pushes out of the body surface as a series of tentacle–like projection called tube feet. It consists of a madreporite, a finely grooved sieve through which water enters, the stone canal which runs from the madreporite to the ring canal around the mouth, five radial canals, lateral canals and the tube feet. Respiration is by dermal branchiae or tube feet. Amebocytes function in respiration, circulation and excretion.

This phylum is divided into Class Asteroidea, the sea stars; Class Ophiuroidea, the brittle stars; Class Echinodermata, the sea urchins; Class Holothuroidea, the sea cucumbers, and Class Crinoidea, the sea lilies.

Phylum Chordata

A living endoskeleton is characteristic of the entire phylum. Two endoskeletons are present: the notochord, present in all members at some time and the vertebral column which replaces the notochord in higher chordates. A ventral heart with a closed circulatory system is present, as well as a hepatic portal system which is specialized for conveying food–laden blood from the digestive system to the liver. Pharyngeal gills are present for the first time. Gill slits or traces are present in the embryos of all chordates. Lungs are a modification of this pharyngeal region. A dorsal, hollow nerve cord is universally present at some stage.

LABORATORY EXERCISE PROCEDURE

Carefully read the discussion and the appropriate chapter(s) in your textbook before attempting the following exercises. Study the figures in the *Photo Atlas for Biology*. When the exercise is complete, submit the report sheets to the instructor for grading.

LABORATORY EXERCISES

I. Phylum Porifera—*Photo Atlas for Biology*, pages 83–85

    A. Look at figures 83a–d and answer the questions in the lab report.

    B. After studying figures 83d and 84c, answer the questions in the lab report.

    C. Look at figures 84a and 85a–c and answer the questions in the lab report.

II. Phylum Cnidaria—*Photo Atlas*, pages 86–90

    A. Draw and label a longitudinal section of a Hydra. Include hypostome, tentacle, gastrovascular cavity, epidermis, gastrodermis, mesoglea and bud.

    B. Draw and label a *Gonionemus medusa* (fig. 88a). Include bell, mesoglea, epidermis, mouth, gonad, and tentacle.

C. Answer the questions in the lab report.

III. Phylum Platyhelminthes—*Photo Atlas*, pages 91–93

    A. Draw and label a planaria. Include head, auricle, eyespot, intestine, pharynx, and pharyngeal pouch.

    B. Draw and label the human liver fluke. Include mouth, oral sucker, pharynx, intestinal ceca, genital pore, ventral sucker, uterus, yolk glands, ovary, seminal receptacle, testes, and excretory pore.

    C. Answer the questions in the lab report.

IV. Phylum Aschelminthes—*Photo Atlas*, pages 94–96

    A. Draw and label a male *Ascaris*. Include anterior end, testis, vas deferens, intestine, seminal vesicle and cloacal opening.

    B. Draw and label a female *Ascaris*. Include anterior end, intestine, vagina, uterus, oviduct, and ovary.

    C. Answer the questions in the lab report.

V. Phylum Mollusca—*Photo Atlas*, pages 97–98

    A. Draw and label a freshwater clam. Include anterior adductor muscle, anterior retractor muscle, labial palps, foot, gills, mantle, auricle of the heart, pericardial sac, posterior adductor muscle, excurrent siphon, and incurrent siphon.

    B. Answer the questions in the lab report.

VI. Phylum Annelida—*Photo Atlas*, pages 99–100

    A. Draw and label the cross section of an earthworm. Include epidermis, circular muscle, longitudinal muscle, cuticle, intestinal wall, typhlosole, dorsal blood vessel, ventral blood vessel, and ventral nerve cord.

    B. Answer the questions in the lab report.

VII. Phylum Arthropoda—*Photo Atlas*, pages 101–104

Answer the questions in the lab report.

VIII. Phylum Echinodermata—*Photo Atlas*, pages 105–107

    A. Draw and label the aboral view of a sea star. Include arm, madreporite, central disks, spine, gonad, ambulacral ridge, digestive gland, digestive gland duct, stone canal, and pyloric stomach.

    B. Answer the questions in the lab report.

SURVEY OF THE ANIMAL KINGDOM

Report Sheet 1

Name _____

Student ID # _____

Campus _____

Date _____

Note: All drawings must be sketches of the pictures in the *Photo Atlas*. Copying drawings from any other source is not acceptable. Credit will not be given for drawings from other sources or for drawings on the wrong page or place on the page.

I. Phylum Porifera

   A. Figures 83a–c

   Discuss the path of water flow through a sponge. Include in your discussion the function of collar cells.

   _____

   _____

   _____

   _____

   _____

   B. Figures 83d and 84c

   What is the function of spicules? _____

   _____

   Spongin fibers? _____

   _____

   C. Figures 84a and 85a–c

   List two types of asexual reproduction in the sponge: _____

   and _____

Survey of the Animal Kingdom　　　　　Student ID # _____
Report Sheet 2

II.　Phylum Cnidaria

　　A.　Draw and label the longitudinal section of a Hydra. Include hypostome, tentacle, gastrovascular cavity, epidermis, gastrodermis, mesoglea and bud.

　　B.　Draw and label a *Gonionemus* medusa. Include bell, mesoglea, epidermis, mouth, gonad, and tentacle.

　　C.　Answer the following questions:

　　　　1.　Distinguish between sexual and asexual reproduction in the hydra. _____
　　　　　　_____
　　　　　　_____

　　　　2.　What is the function of nematocysts? _____
　　　　　　_____

128　Exercise 14

Survey of the Animal Kingdom　　　　　Student ID # _____
Report Sheet 3

   3. Discuss the food–getting of hydra. Include the function of tentacles and the significance of a single opening into the gastrovascular cavity.

      _____

      _____

      _____

   4. What is the difference between a polyp and a medusa? _____

      _____

      _____

   5. List the four different types of Cnidaria

      a. _____

      b. _____

      c. _____

      d. _____

III. Phylum Platyhelminthes

   A. Draw and label a planaria. Include head, auricle, eyespot, intestine, pharynx, and pharyngeal pouch.

Survey of the Animal Kingdom Student ID # _____
Report Sheet 4

    B.   Draw and label the human liver fluke. Include mouth, oral sucker, pharynx, intestinal ceca, genital pore, ventral sucker, uterus, yolk glands, ovary, seminal receptacle, testes, and excretory pore.

    C.   Answer the following questions:

        1.   What is the function of the hooks and suckers on the scolex of the tapeworm?

_____

_____

_____

        2.   What does a gravid proglottid contain? _____

_____

IV.   Phylum Aschelminthes

    A.   Draw and label a male *Ascaris*. Include anterior end, testis, vas deferens, intestine, seminal vesicle and cloacal opening.

130   Exercise 14

Survey of the Animal Kingdom Student ID # _____
Report Sheet 5

    B.   Draw and label a female *Ascaris*. Include anterior end, intestine, vagina, uterus, oviduct and ovary.

    C.   Answer the following questions:

        1.   What is the easiest way to tell a male *Ascaris* from a female? _____

            _____

            _____

        2.   The trichina worm encysts in what type of muscle?_____

            _____

V.   Phylum Mollusca

    A.   Draw and label a freshwater clam. Include anterior adductor muscle, anterior retractor muscle, labial palps, foot, gills, mantle, auricle of the heart, pericardial sac, posterior adductor muscle, excurrent siphon, and incurrent siphon.

Survey of the Animal Kingdom   Student ID # _____
Report Sheet 6

B. Answer the following questions:

1. What is the function of the foot? _____
   _____
   _____

2. What is the function of the mantle? _____
   _____
   _____

3. What structures are found within the foot? _____
   _____
   _____

4. What are the functions of the incurrent and excurrent canals? _____
   _____
   _____
   _____

5. What is the function of the digestive gland? _____
   _____
   _____

6. List three types of mollusks seen in the *Photo Atlas*.

   a. _____

   b. _____

   c. _____

132   Exercise 14

Survey of the Animal Kingdom Student ID # _____
Report Sheet 7

VI. Phylum Annelida

    A. Draw and label the cross section of an earthworm. Include epidermis, circular muscle, longitudinal muscle, cuticle, intestinal wall, typhlosole, dorsal blood vessel, ventral blood vessel, and ventral nerve cord.

    B. Answer the following questions:

       1. List the parts of the earthworm digestive system.

          a. _____ e. _____

          b. _____ f. _____

          c. _____ g. _____

          d. _____

       2. What is the function of the typhlosole? _____

          _____

       3. How does the leech attach itself to its host? _____

          _____

       4. How were leeches used at the turn of the century? _____

          _____

Survey of the Animal Kingdom Student ID # _____
Report Sheet 8

VII. Phylum Arthropoda

    A. List the two major sections of the crayfish.

        1. _____

        2. _____

    B. Answer the following questions:

        1. How many walking legs and swimmerets do crayfish have?

           a. Walking legs _____

           b. Swimmerets _____

        2. What do crayfish use for respiration? _____

        3. How does the grasshopper differ from the crayfish in terms of major body sections?

           _____

           _____

           _____

        4. Based on figures 104a and 104b in the *Photo Atlas*, what is the major difference between millipedes and centipedes? (Hint: Look at attachments.)

           _____

           _____

        5. List the seven structures or physiological patterns that make the arthropods so successful.

           a. _____   e. _____

           b. _____   f. _____

           c. _____   g. _____

           d. _____

        6. Which class of arthropods is the most successful? _____

           _____

Survey of the Animal Kingdom Student ID # _____
Report Sheet 9

VIII. Phylum Echinodermata

  A. Draw and label the aboral view of a sea star. Include arm, madreporite, central disc, spine, gonad, ambulacral ridge, digestive gland, digestive gland duct, stone canal, and pyloric stomach.

  B. Answer the following questions:

   1. Discuss the water–vascular system of the sea star including the terms tube feet, madreporite, stone canal, and ring canal.

   _____
   _____
   _____
   _____
   _____
   _____

   2. Does the sea cucumber exhibit radial symmetry? _____ How can you tell?

   _____

Survey of the Animal Kingdom   Student ID # _____
Report Sheet 10

3. List six different echinoderms as seen in the *Photo Atlas*.

   a. _____   d. _____

   b. _____   e. _____

   c. _____   f. _____

4. List six major characteristics of mollusks.

   a. _____   d. _____

   b. _____   e. _____

   c. _____   f. _____

IX. Analysis and Conclusions

   A. Answer the following questions:

   1. Why are the sponges considered to be very primitive animals compared to the echinoderms when both phyla exhibit radial symmetry, a more primitive type of symmetry?

   _____

   _____

   _____

   2. What is the advantage of the medusa over the attached polyp in the Cnidaria?

   _____

   _____

   _____

   3. When do three distinct germ layers appear in the animal kingdom?_____

   _____

   4. When does the "hint" of a primitive nervous system appear? _____

   _____

Survey of the Animal Kingdom Student ID # _____
Report Sheet 11

5. After looking at the liver fluke and tapeworm, explain the absence of a major digestive system.

   _____

   _____

   _____

6. How can the life cycle of the tapeworm be broken? _____

   _____

7. How can the life cycle of the liver fluke be broken? _____

   _____

8. Why does *Ascaris* contain mostly a reproductive system? _____

   _____

   _____

9. When does most of the damage occur when infested with roundworms? _____

   _____

   _____

10. Why are poorly developed countries more prone to infestations of roundworms?

    _____

    _____

11. When does a complete digestive system appear in the Animal Kingdom?

    _____

12. Which phylum has fossil records as far back as the Cambrian Period?

    _____

13. With Phylum Annelida, the excretory system has made what advantage?

    _____

Survey of the Animal Kingdom   **137**

Survey of the Animal Kingdom　　　Student ID # _____
Report Sheet 12

14. When does a closed circulatory system appear in the Animal Kingdom?

    _____

15. List four advances that represent characteristics of higher forms of animals.

    a. _____   c. _____

    b. _____   d. _____

16. Which phylum has reached the peak of invertebrate evolution? _____

    _____

17. Why is Insecta the most successful class of Arthropods? _____

    _____

    _____

18. List four phyla that have reached the highest level of organization.

    a. _____   c. _____

    b. _____   d. _____

19. What two structures are present in all Chordates at some time in their development?

    a. _____   b. _____

20. How do the Arthropods differ from the Chordates in terms of a skeleton?

    _____

    _____

    Which is more efficient and why? _____

    _____

    _____

138　Exercise 14

Survey of the Animal Kingdom
Report Sheet 13

Student ID # _____

21. Of what benefit are a closed circulatory system, a closed digestive system, an excretory system and a nervous system to higher animals?

_____

_____

_____

_____

_____

_____

15 PLANTS: TISSUES, NUTRITION, AND TRANSPORT

LESSON OBJECTIVES

Upon completion of this laboratory exercise the student will be able to:

1. Describe the parts of a flowering plant.

2. Define the types of dermal tissues, ground tissues, and vascular tissues in the flowering plants and give functions for each.

3. Give general functions of roots, stems and leaves.

4. Distinguish between monocot and dicot roots, stems and leaves.

5. Distinguish between spring and fall wood in dicot stems.

6. Outline the path of water in plants.

7. Describe the structure of a leaf in terms of gas exchange and the prevention of water loss.

MATERIALS NEEDED

1. Laboratory Manual
2. Textbook
3. Colored pencils
4. *Photo Atlas for Biology*
5. Fresh celery stalk
6. Container to hold the celery stalk upright
7. Food coloring, preferably red (but other colors can be used)

PREPARATION

Read the discussion which follows carefully before attempting to complete this exercise. Also read the appropriate chapter(s) in your textbook.

DISCUSSION

All life on this planet is dependent on green plants which are the primary producers. Plants occupy this position of importance because they have the ability to take energy from sunlight, convert the energy to chemical energy, and store energy in the form of sugar. The flowering plants carry on life activities with the specialized structures called roots, stems, leaves and reproductive structures. The reproductive structures will be covered in another exercise.

All parts of the plant have three tissue systems: dermal, vascular and ground. The **dermal tissue** system provides a covering for the plant body. It consists of **epidermal cells**, cells which secrete a waxy layer, the **cuticle**, which restricts water loss from the plant surface in above-ground plant structures and epidermal outgrowths, **trichomes**, with specialized functions such as the root hairs which increase the surface area of the roots for water absorption.

The **vascular tissue system** is responsible for conduction of substances in the plant such as water, dissolved minerals and food. The vascular system is composed of **xylem** and **phloem**. The primary function of xylem is to conduct water and dissolved minerals from the roots to the stems and leaves. A secondary function is structural support. Food is conducted through the plant by phloem cells.

Flowering plants are also divided into two groups based on the number of cotyledons (primitive fleshy leaves located inside the seed), flower parts, leaf venation and the arrangement of their vascular bundles (xylem and phloem). The **dicots** have 2 cotyledons, flower parts in multiples of 4 or 5, netted venation in the leaves and distinct circular bands of xylem and phloem in the stem or trunk. **Monocots** have one cotyledon, flower parts in multiples of 3, parallel venation in leaves, and the xylem and phloem are scattered throughout the stem and root.

The rest of the plant body is composed of the **ground tissue system** with various cell types and functions. Chloroplasts are found in the ground tissue system as are storage cells for starch.

The main functions of **roots** are anchorage, absorption, conduction and storage. Primary roots have an epidermis, cortex, endodermis, pericycle, xylem and phloem. The **epidermis** protects the root. The **cortex** contains storage tissue. The endodermis controls water uptake by the root. Branch roots originate in the **pericycle**. Monocot roots often have a pith while dicot roots have a **vascular cambrium**, secondary xylem which becomes wood and secondary phloem which becomes inner bark.

Stems function in support, internal transport of materials, and production of new stem tissue. Stems with primary growth have an epidermis, vascular tissue, cortex and pith. The **epidermis** functions as protection for the stem. The **vascular tissue** is xylem and phloem. The **cortex** is tissue just inside the epidermis and **pith** is tissue at the center of the plant stem. Dicot stems have a distinct cortex and pith. The vascular bundles of xylem and phloem are arranged in a circle. The monocot stems have ground tissue instead of a distinct cortex and pith. Their vascular bundles are scattered throughout the ground tissue.

The main function of leaves is photosynthesis. The **leaf** consists of epidermis, mesophyll and vascular tissue. The leaf is generally flat with an upper and lower epidermis. The epidermis has openings called **stomata** which allow the exchange of gases into and out of the leaf. The photosynthetic tissue of the leaf is the **mesophyll**, tissue containing chloroplasts. The vascular tissue is located in veins that extend through the mesophyll. The arrangement of veins in leaves, the **venation**, is used to distinguish between a monocot and dicot. The monocots have veins which are parallel, such as in the grasses, corn and iris. The dicots, such as maple and oak, have veins which form a net-like pattern called net venation.

While food in the form of starch is produced during photosynthesis in chloroplasts in leaves and transported from leaves to roots through phloem cells, water transport occurs in xylem cells from root to leaf. **Root pressure**, caused by the movement of water into the root from the soil, explains the rise of water in small plants. In large plants, **tension-cohesion** causes the rise of water. The evaporation pull of **transpiration** in leaves produces a tension at the top of the plant. A solid column of water is pulled up through the xylem as a result of the cohesive and adhesive properties of water.

LABORATORY EXERCISE PROCEDURE

Carefully read the procedure before completing this exercise. Then complete the lab report and submit it to your instructor for grading.

LABORATORY EXERCISE

I–V. Use the *Photo Atlas for Biology* and your textbook to answer the questions.

VI. Transport in a Celery Stalk

    A. Objective: To observe water transport in a plant

    B. Materials

        1. Fresh celery stalk with leaves
        2. Container of water large enough to hold the celery stalk upright
        3. Red food coloring (some other color could also be used)

    C. Procedure

        1. Cut off the end of the celery stalk while holding the stalk under water.
        2. Add food coloring to the water until the water has a *distinct* color—red or blue is best.
        3. Leave the stalk in the water for several days.
        4. Observe the appearance of the stalk and the leaves several times a day.
        5. Record your observations in the lab report.

PLANTS: TISSUES, NUTRITION, AND TRANSPORT Report Sheet 1

Name _____

Student ID # _____

Campus _____

Date _____

I. Monocytes and Dicots

    A. Describe a monocot _____

    _____

    _____

    B. Describe a dicot _____

    _____

    _____

    C. Draw a monocot and a dicot seed. Label the plumule cotyledons in both. Refer of pg. 66, fig. 66c and pg. 67, fig. 67d, in the *Photo Atlas*.

                Monocot   Dicot

Plants: Tissues, Nutrition, and Transport   Student ID # _____
Report Sheet 2

    D. Draw a section of a monocot and dicot leaf. (Your drawing must represent what you see in the *Photo Atlas*, figs. 82a and 82b, pg. 82, for credit.)

                Monocot   Dicot

    E. Compare the monocot and dicot stems (use figs. 77a and 77b, pg. 77, in the *Photo Atlas* for reference).

        Monocot stem _____

        _____

        _____

        Dicot stem _____

        _____

        _____

        1. What is cambium? _____

        2. What are annual rings? _____

II. The Leaf

    A. In fig. 82a, pg. 82 of the *Photo Atlas*, why are the palisades layer and spongy layer of the leaf called the working layer?

        _____

        _____

        _____

Plants: Tissues, Nutrition, and Transport   Student ID # _____
Report Sheet 3

B. Looking at fig. 81a, pg. 81, how many layers of cells make up the epidermis? _____

C. After looking at figs. 81b, 81c, and 81e, pg. 81, sketch a stomate.

D. Suggest a mechanism for controlling water loss in the leaf and gas exchange in the leaf.

_____

_____

_____

_____

III. The Stem

A. Draw a pie–shaped wedge of the corn stem. Label the epidermis, pith, and vascular bundle. In one vascular bundle label the xylem and phloem (use figs. 77c and d, pg. 77 in the *Photo Atlas* for help).

146   Exercise 15

Plants: Tissues, Nutrition, and Transport   Student ID # _____
Report Sheet 4

   B. Draw a pie–shaped wedge of a dicot stem (figs. 77a & b, pg. 77). Label the epidermis, ortex, pith, xylem, phloem and cambrium.

   C. After looking at fig. 77a, pg. 77, and fig. 79a, pg. 79, discuss the differences in a non-woody dicot and a woody dicot stem. Be specific.

_____

_____

_____

III. The Root

   A. Sketch a pie–shaped wedge of the root cross–section in figs. 75a & b, pg. 75. Label the central cylinder, the cortex, epidermis, xylem and phloem.

Plants: Tissues, Nutrition, and Transport   Student ID # _____
Report Sheet 5

B. What is found in the cortex cells? _____
_____
_____

C. What is the function of the root cap? _____
_____
_____

D. The zone of elongation? _____
_____
_____

IV. Water Transport

A. Describe the flow of water from outside the root, through the stem to the leaf. Be specific but concise. Refer to your textbook for help.
_____
_____
_____
_____
_____

B. Describe the appearance of the celery stalk in the colored water. _____
_____
_____
_____

Explain what you see. _____
_____
_____

Exercise 15

Plants: Tissues, Nutrition, and Transport   Student ID # _____
Report Sheet 6

    C.   Describe the appearance of the leaves in the colored water. _____
_____
_____

        Explain what you see. _____
_____
_____

V.   Analysis and Conclusion

    A.   Explain how a root increases in length. _____
_____
_____

    B.   How does the arrangement of vascular bundles differ in monocot and dicot stems?
_____
_____
_____

    C.   How does the location of the cambium layer make it possible for stems to grow in diameter?
_____
_____
_____

    D.   How does the shape of a leaf help it to use sunlight efficiently? _____
_____
_____
_____

Plants: Tissues, Nutrition, and Transport   Student ID # _____
Report Sheet 7

E. In which layer of the mesophyll of a leaf does most photosynthesis occur? _____

How can you determine this? _____

_____

F. Name the leaf structure that helps to conserve water. _____

_____

_____

150   Exercise 15

16 PLANTS: REPRODUCTION AND DEVELOPMENT

LESSON OBJECTIVES

Upon completion of this laboratory exercise the student will be able to:

1. Describe a generalized plant life cycle, identifying haploid and diploid cells.

2. Discuss alternation of generations in plants; including mosses, ferns, gymnosperms and angiosperms.

3. Label the parts of a perfect flower.

4. Label the parts of a seed.

5. List the major types of fruits.

6. Give functions of the seed and fruit.

7. Outline the process of germination and early plant growth.

MATERIALS NEEDED

1. Laboratory Manual
2. Textbook
3. *Photo Atlas for Biology*
4. Corn seed soaked in water overnight
5. Lima bean seed soaked in water overnight
6. Small knife
7. Various fruits such as:
   tomato
   cherry or peach or olive with "seed"
   apple
   fresh orange or lemon or grapefruit
   fresh strawberry
   fresh string bean
   squash
   pineapple
8. Hand lens*

   *This item is found in your lab kit.

PREPARATION

Read the discussion which follows carefully before attempting to complete this exercise. Also read the appropriate chapter(s) in your textbook.

DISCUSSION

All plants exhibit **alternation of generations**; that is, they spend part of their life cycle in the haploid stage and half of their life cycle in the diploid stage. The haploid portion of the life cycle is called the **gametophyte generation** because it gives rise to gametes by the process of mitosis. The diploid portion of the life cycle is the **sporophyte generation**, which produces spores immediately following meiosis.

The gametophytic plant produces **antheridia** and **archegonia**. Sperm travel to the archegonia in a variety of ways and one sperm fuses with the egg. This process, known as **fertilization**, results in a fertilized egg, **or zygote**. The zygote is diploid (2N) and is the first cell in the sporophyte generation. The zygote divides by mitosis and develops into a multicellular embryo. Eventually the embryo develops into the sporophyte plant. The sporophyte plant has specialized cells which divide by meiosis, forming haploid spores. These spores represent the first stage in the gametophyte generation.

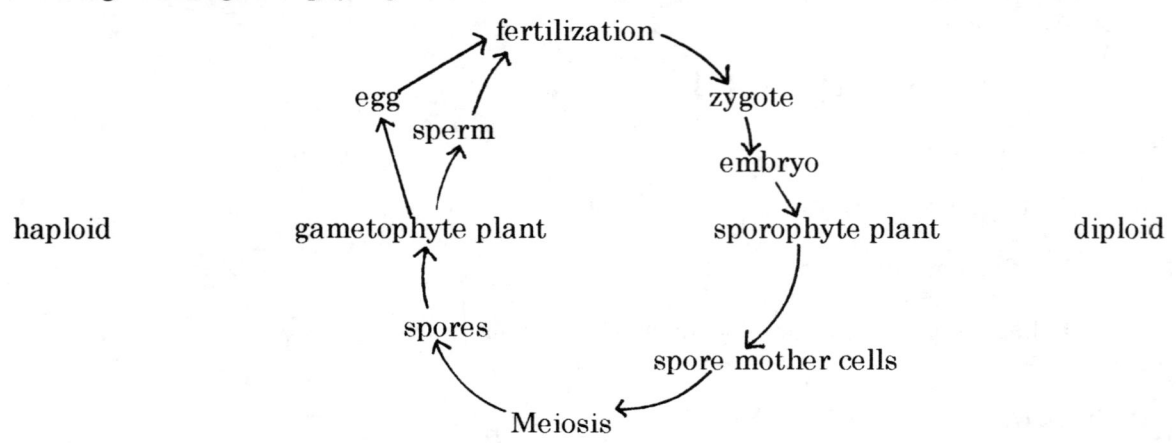

```
                    fertilization
            egg                    zygote
              ↖  sperm           ↙
                                embryo
                                  ↓
haploid    gametophyte plant    sporophyte plant      diploid
              ↑
            spores
                  ↖  spore mother cells
                     Meiosis
```

The gametophyte generation is the more dominant generation in primitive plants such as the moss and liverwort. The sporophyte generation is the dominant generation in the ferns, angiosperms and gymnosperms.

The primary means of reproduction and dispersal for the most successful plants is by **seeds**, which develop from the female gametophyte and the tissues associated with it. Seeds contain a multicellular well–developed young plant with embryonic root, stem and leaves. After germination, the embryo within the seed is nourished by food stored inside the seed. Both angiosperms and gymnosperms produce seeds. The gymnosperms produce seeds that are totally exposed or borne on the scales of cones. Angiosperms produce their seed within a fruit.

The organ of sexual reproduction in angiosperms is the **flower**. Flowers consist of sepals, petals, stamens and carpels. A specialized stalk, the **pedicel**, gives rise to the flower. The **receptacle**, located at the apex of the pedicel, is an enlarged area where the floral parts are attached. At the base of the flower are the **sepals** which together form the **calyx**. Proceeding toward the center of the flower from the sepals, the **corolla** is encountered next. The corolla is composed of **petals**. The **stamens** are next. Each stamen consists of a slender elongated stalk, the **filament**, with an enlarged end, the **anther**. The anther contains the **pollen grains**. Located in the center is the **pistil** or **pistils**. The uppermost section of the pistil is flattened to receive the pollen grains. A stalklike portion, the **style**, extends from the stigma to the ovary. The **ovary** is at the base of the flower and contains the **ovules** or eggs. After fertilization of the eggs within the ovary, each ovule develops into a **seed** and the ovary develops into a **fruit**.

Double fertilization is unique to angiosperms. It occurs because the angiosperms are heterosporous, producing microspores and megaspores. Each ovule within the ovary contains a megaspore mother cell which undergoes meiosis, producing 4 haploid megaspores. One of these haploid megaspores develops into the female gametophyte, the **embryo sac**. The embryo sac contains 8 nuclei, including one egg and 2 polar nuclei. The egg and polar nuclei are involved in fertilization. The anther contains microspore mother cells which undergo meiosis to form haploid microspores. The microspores develop into the male gametophyte, pollen grains. Pollen is transferred to the stigma of the carpel. The pollen grains, if compatible with the female plant, grow tubes through the style and into the ovary. A cell within each pollen grain divides to form 2 sperm nuclei. Both sperm nuclei are involved in fertilization. One sperm nucleus fuses with the egg, forming the diploid zygote that will develop into the embryo plant in the seed. The other sperm nucleus fuses with the 2 haploid polar nuclei, forming a triploid (3N) cell that develops into endosperm in the seed. The **endosperm** is the main source of food for monocot plants; in dicots, the endosperm is absorbed by the embryo, which stores food in its cotyledons. As the seed develops, the ovary wall surrounding it enlarges and develops into a fruit. Fruits protect the developing seeds from desiccation during maturation and often aid in seed dispersal.

The seed consists of a **seed coat**, **embryo** and **endosperm**. The seed of a monocot has one cotyledon. A monocot embryo consists of three parts, the **cotyledon**, the **epicotyl** and the **hypocotyl**. The cotyledon absorbs food from the endosperm and transfers it to the embryo during early growth of the seedling. The epicotyl grows upward from the cotyledon and becomes the shoot above ground. Inside the shoot are the tissues that will later become leaves and stem. The hypocotyl grows downward from the cotyledon. Its tip is the **radicle,** the embryonic root.

The dicot seed has 2 cotyledons. Each seed is covered by a seed coat. The bean seed is attached to its pod at the **hilum**. At one end of the hilum is a small pore called the **micropyle**, where the pollen tube entered the ovule of the flower. The micropyle is the only place where water can enter the seed. In the dicot the inner seed is made up entirely of the embryo. The **embryo** consists of the 2 large cotyledons, the epicotyl and hypocotyl. The **cotyledons** contain stored food. The **plumule** in the bean seed consists of the tiny leaves. They are part of the **epicotyl**, covering its tip. The plumule will become the first leaves.

The **fruit** is the ripened ovary surrounding one or more seeds. The three major types of fruit are **simple fruits**, **aggregate fruits** and **multiple fruits**. A simple fruit develops from a single pistil in a flower. A simple fruit may be either dry like peas and string beans or fleshy like the berry, pome and drupe. In a **pome**, the receptacle develops into the fleshy part of the fruit as it grows and surrounds the ovary. The ovary develops into the inner core which is the true fruit. Examples are apples and pears. A **drupe** is also called a stone fruit. All parts of the fruit form from the ovary. The skin and fleshy layer are the two outer layers of the ovary wall. The hard part of the stone is the inner layer of the ovary wall. Cherries, peaches, plums and olives are drupes. A true **berry** is a fleshy, enlarged ovary. The seeds are embedded in the flesh. A thin skin covers the fruit. Grapes, tomatoes and blueberries are examples of berries. A **pepo** is a type of berry with a firm rind which does not separate from the flesh. Cucumbers, squash, melons and pumpkins are pepos. A **hesperidium** is a type of berry with a leathery rind like citrus fruits.

Aggregate fruits develop from single flowers that have many separate pistils. The pistils develop into a number of small fruits that mature together on a single receptacle. The raspberry, blackberry and strawberry are examples of aggregate fruits.

A **multiple fruit** forms from the ovaries of many flowers closely clustered on a central stalk. The pineapple and mulberry are examples of aggregate fruits.

LABORATORY EXERCISE PROCEDURE

Carefully read the procedure before completing this exercise. Then complete the lab reports and submit them to your instructor for grading.

LABORATORY EXERCISE

I. Flower

   A. Use the *Photo Atlas for Biology*, pages 59–61

   B. Procedure

   1. Draw and label a generalized picture of a monocot and a dicot flower, showing filament, anther, stigma, style and ovary.
   2. Answer the questions in the lab report.

II. Seed

   A. Corn Grain—Using the hand lens, observe a corn grain soaked in water overnight. Sketch it and label the embryo and the endosperm.

   B. Lima Bean Seed

   1. Examine the soaked lima bean seed and observe the hilum. Use the hand lens to find the micropyle.
   2. Sketch the bean seed with the hilum facing you. Label the hilum, seedcoat and micropyle.
   3. Peel the seed coat from the seed very carefully. Observe the plant embryo. Be careful not to break off any part of the embryo. Carefully open the embryo to expose the cotyledons, hypocotyl, epicotyl and the plumule. Sketch the embryo, labeling all of the structures listed above.
   4. Answer the questions in the lab report.

III. Fruit

   A. Simple fruit

   1. Examine a fresh string bean. Sketch the pod labeling the peduncle, receptacle, sepals, and seeds.
   2. Study an apple. Cut it in half lengthwise through the center of the core. Sketch the apple labeling pedicel, receptacle, ovary and sepals.
   3. Break open the stone inside a drupe such as the cherry. Describe what you observe.
   4. Cut a tomato crosswise through its center. Note the many seeds embedded in the pulpy flesh. Can you peel off the skin?
   5. Cut an orange crosswise through its center. Examine the inner structure.
   6. Examine a strawberry and determine where the seeds are located. Approximately how many fruits make up the strawberry?
   7. Examine a cut section of a pineapple. Note the central core. it was a central stalk that bore a number of separate flowers. Observe the outside of the fruit. Each rounded section came from one flower. Answer question 9 in the lab report.

PLANTS: REPRODUCTION AND DEVELOPMENT

Report Sheet 1

Name _____

Student ID # _____

Campus _____

Date _____

I. The Flower

   A. Draw and label a generalized picture of a monocot and a dicot flow, showing the filament, anther, stigma, style and ovary for each. Refer to pages 59 and 60 in the *Photo Atlas*.

   Monocot   Dicot

   1. Number of sepals for a monocot? _____ Dicot? _____

      Number of petals for a monocot? _____ Dicot? _____

   B. Answer the following questions:

      1. Where are the male gametes produced? _____

      2. How do the male gametes reach the female gametes? _____
      _____

      3. Where are the female gametes produced? _____

      4. Describe a hypogynous flower arrangement _____
      _____

Plants: Reproduction and Development   Student ID # _____
Report Sheet 2

5. What does perigynous mean? _____
   _____

6. What does epigynous mean? _____
   _____

7. What does inflorescence mean in terms of a flower? _____
   _____

II. Seed

   A. Sketch the corn grain—label the embryo and endosperm

   B. Lima Bean

      1. Sketch the lima bean seed—label the hilum, micropyle and seed coat

      2. Sketch the bean embryo—label the cotyledons, hypocotyl, epicotyl and plumule. Refer to figures 66c and 67a, pages 66 and 67 for help.

Exercise 16

Plants: Reproduction and Development   Student ID # _____
Report Sheet 3

    3. How many leaves make up the plumule? _____

    4. Can you see veins in the leaves of the plumule? _____

III. Fruit

    A. Sketch a string bean—label the peduncle, receptacle, sepals and mature seeds. Refer to figure 66b, page 66.

    B. Sketch an apple—label the pedicel, receptacle, ovary and sepals. Refer to figure 68c, page 68 in the *Photo Atlas*.

    C. Give a brief description of what you observe inside the stone of a drupe.

    _____

    _____

    _____

    D. Can the skin of a true berry be pulled off? _____

Plants: Reproduction and Development   Student ID # _____
Report Sheet 4

E. Describe the inner structure of an orange. _____
_____
_____
_____

D. Does the rind separate easily from the flesh in a hesperidium? _____
_____

E. Where are the seeds located in a strawberry? _____
_____

F. How many fruits make up a strawberry? _____
_____

G. How many flowers were clustered to form the section of pineapple? _____
_____

IV. Analysis & Conclusions

A. Some flowers are wind–pollinated, while others are pollinated by insects. What flower parts would a wind–pollinated flower have in common with an animal pollinated flower?
_____

What parts would be different? _____

Explain _____
_____
_____

B. Explain why pollen allergies would be blamed on wind–pollinated plants rather than on animal–pollinated plants.
_____
_____
_____

158   Exercise 16

Plants: Reproduction and Development   Student ID # _____
Report Sheet 5

C. What part of a seed provides us with nutrients? _____
   _____

D. Some fruits are brightly colored and some taste sweet. How would these qualities aid in the dispersal of seeds?

   _____
   _____
   _____

E. The raspberry is an aggregrate fruit. It is made up of many simple fruits attached to a receptacle. Where are the raspberry seeds?

   _____
   _____

   Be specific_____
   _____
   _____

17 ANIMAL TISSUES

LESSON OBJECTIVES

Histology is the study of tissues. A tissue, by definition, is an aggregation of cells working together to perform a specialized activity. At the completion of this laboratory, the student will be able to:

1. Define a tissue.

2. Make sketches and drawings of specified tissues labeling indicated structures from the *Photo Atlas*.

3. Classify the tissues of the human body into four major types.

4. Describe the structure of epithelial, connective, muscular, and nervous tissues and their location and major function in the human body.

MATERIALS NEEDED

1. Laboratory manual
2. Pencil
3. *Photo Atlas for Biology*
4. Textbook

PREPARATION

Read the following discussion carefully before attempting the laboratory exercises. Also read the appropriate chapter(s) in your textbook.

DISCUSSION

Previous lessons have examined examples of various plant and animal cells. It has been noted that each cell is a highly organized unit composed of numerous organelles. In complex organisms, such as the human, cells do not function independently of one another but are organized into tissues. A **tissue**, by definition, is an aggregation of cells working together to perform some specialized activity. The field of biology concerned primarily with the study of tissues is termed **histology**. Histologists recognize four major types of tissues according to their structure and functions. These include epithelial tissue, connective tissue, muscular tissue and nervous tissue.

Epithelial tissue covers the surface of the body, lines tubes (such as the gastrointestinal tract) and cavities. Epithelial tissue also forms the secretory cells of glands. The functions of epithelial tissue include absorption, secretion, diffusion, filtration and protection.

Connective tissue connects muscle to bone, furnishes nourishment to overlying epithelial tissues, provides support for other tissues and organs, and binds organs together.

Muscular tissue is responsible for movement. This includes movement of the extremities, head and torso, as well as internal movement to perform functions such as moving food through the digestive system and pumping blood.

Nervous tissue receives and transmits stimuli throughout the body.

Epithelial Tissue

Epithelial tissue can be divided into two types: that which covers and lines and that which constitutes the secreting portion of glands. Covering or lining type epithelium may consist of a single layer of cells or it may be several layers thick. The single layer type is called **simple epithelium** and is found in areas such as the small intestine. It functions in absorption and filtration. If the epithelial tissue is in an area of maximum wear or tear, such as the outer skin, and is layered, it is termed **stratified epithelium**.

The shapes of the individual cells composing epithelial tissue are also used to further classify the tissues. Flat, close fitting cells with large, centrally located nuclei are termed **squamous cells**. These are found lining the air sacs of the lungs, in the crystalline lens of the eye, in the glomerular capsule of the kidneys, and lining the eardrum. These cells function in absorption and filtration.

Single layered, cube–shaped cells with centrally located nuclei are termed **cuboidal cells** and function in absorption and secretion. Cuboidal epithelium is located in kidney tubules, lining the inner surface of the cornea and lens of the eye, and covering the surface of the ovary.

Elongated, rectangular shaped cells with their nuclei at the base of the cells are termed **columnar epithelium** and function in protection, secretion, and absorption. Some columnar cells have finger–like projections on the top known as **cilia** which facilitate movement across the cell surface. **Ciliated columnar epithelium** lines the upper respiratory tract, Fallopian tubes and ducts of the testes. **Nonciliated columnar epithelium** lines the stomach, small and large intestines, gallbladder and digestive glands. Two types of cells may be present: **goblet cells** which are the mucus secreting specialists of the body and columnar cells which are the body's absorption specialists.

Stratified epithelium as either a covering or a lining tissue also occurs in various cell shapes including squamous, cuboidal, columnar and transitional. **Stratified squamous** tissue is found lining the mouth, esophagus, and vagina where it serves as a protective layer of tissue. The outer layer of human skin is composed of **stratified squamous keratin–rich tissue** (keratin is a protein which aids in waterproofing and helps toughen skin), and its major function in this location is to protect the body from foreign invasion.

The ducts of adult sweat glands are lined with **stratified cuboidal epithelium** while the male urethra and some excretory ducts are lined with **stratified columnar epithelium**. Both types give protection while the columnar cells function in secretion. **Transitional epithelium**, a stratified epithelium, lines the urinary tract and permits distention as these cells can change shape. This tissue resembles stratified squamous epithelium but with large, rounded, superficial cells.

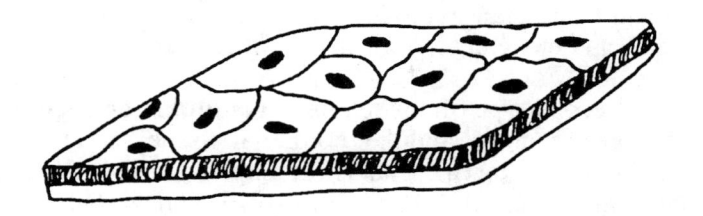

FIGURE 1

Simple Squamous Epithelium

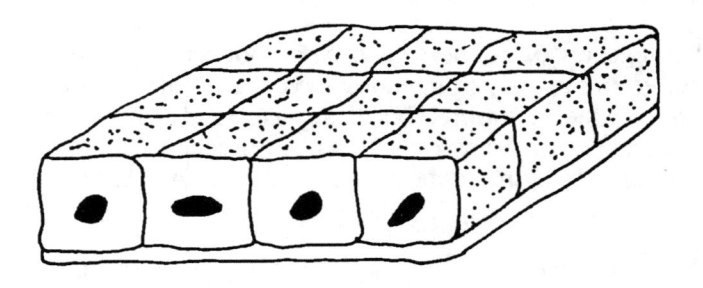

FIGURE 2

Simple Cuboidal Epithelium

— Cilia

— Goblet Cell

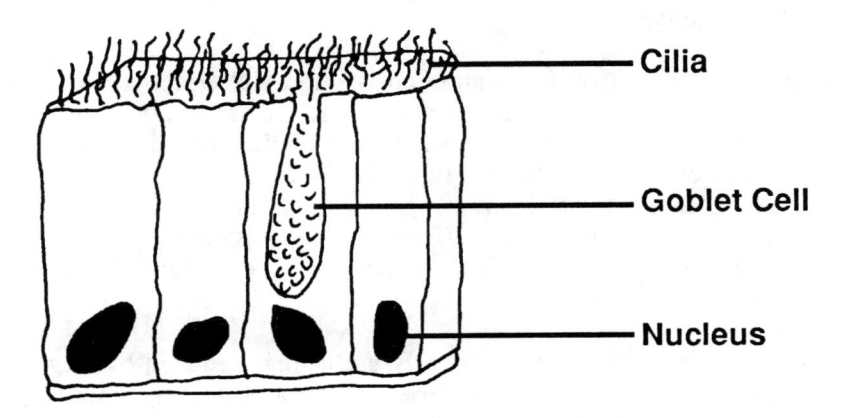

— Nucleus

FIGURE 3

Ciliated Columnar Epithelium

Animal Tissues

Pseudostratified epithelium lines the excretory ducts of some large glands, the male urethra, respiratory passages, and some male reproductive ducts. These cells function in secretion, as they contain goblet cells, and movement of mucus by ciliary action.

In terms of function, the second type of epithelium is **glandular epithelium** which can also be subdivided into two types. Some glandular epithelial tissue secretes products into ducts. Glands of this type are termed exocrine glands. Examples are the sweat, oil and wax glands of the skin; mammary glands; and the liver and salivary glands. Other glandular epithelium secretes products directly into the blood stream. Glands of this type are termed endocrine glands. Examples are the pituitary and adrenal glands.

Connective Tissue

Connective tissue is the most abundant tissue in the human body. This tissue functions in protection and support. It also binds various organs together. Embryonal connective tissue is found only in a developing embryo up to about two months after fertilization has occurred. For this reason, we will only examine adult connective tissue.

There are several categories of connective tissue: connective tissue proper, cartilage, bone, and blood. Connective tissue proper includes loose or areolar, adipose, collagenous, elastic and reticular. **Loose** or **areolar connective tissue** is found in the subcutaneous layers of the skin, nerves, body organs, blood vessels and in the mucus membranes. It consists of fibers and cells which provide strength, support and elasticity. Antibodies and anticoagulants are also produced by this tissue.

Adipose tissue is found in the subcutaneous skin layer, in the marrow of long bones, as padding around joints, and around the kidneys and heart muscle. this tissue is specialized to store fat. Fat serves as reserve food, helps maintain body heat, and protects and supports body organs. The stored fat occupies a large portion of the adipose tissue cell, pushing the nucleus toward the edge so that each cell resembles a ring.

The tendons and ligaments are composed of **collagenous connective tissue** which provides strong attachment between structures. **Elastic connective tissue** permits stretching and is found in the lungs, bronchial tube, larynx, vocal cords, trachea, arterial walls, and between the vertebrae. **Reticular connective tissue** is a network of fibers found in such organs as the liver, spleen and the lymph nodes.

Cartilage is also connective tissue and consists of a gelatinous matrix with collagenous and elastic fibers woven throughout. Its primary function is support; and histologists recognize three types: hyaline, elastic, and fibrocartilage. **Hyaline cartilage** is perhaps best known as gristle. It is found on the nose, larynx, trachea, and ends of long bones. In addition to support, it aids in joint moment. **Elastic cartilage** is found in the auricle of the external ear, the epiglottis and in parts of the larynx. **Fibrocartilage** is found between the vertebrae in the spinal column and in the pubic symphysis where the inferior rami of the pelvis join to form the pubic arch.

Bone tissue, another type of connective tissue, is characterized by large amounts of ossified (hardened) matrix containing cells, arteries, veins and nerves. Microscopic examination of bone tissue reveals hundreds of concentric patterns known as **Haversian canal systems**. The center of each pattern is dominated by a Haversian canal which is surrounded by concentric layers of ossified matrix. The concentric patterns surrounding the Haversian canal contain small cavities termed **lacunae** which house the osteocytes (the living part of bone). Small canals, the **canaliculi**, connect the osteocytes with the blood vessels of the Haversian

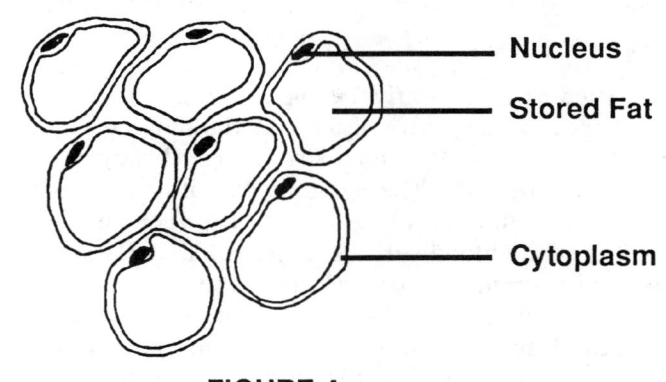

FIGURE 4

Adipose Tissue

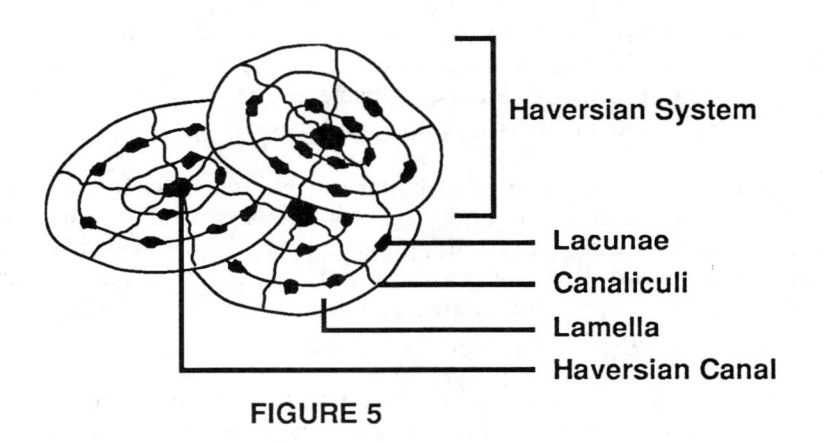

- Nucleus
- Stored Fat
- Cytoplasm

FIGURE 5

Bone Tissue

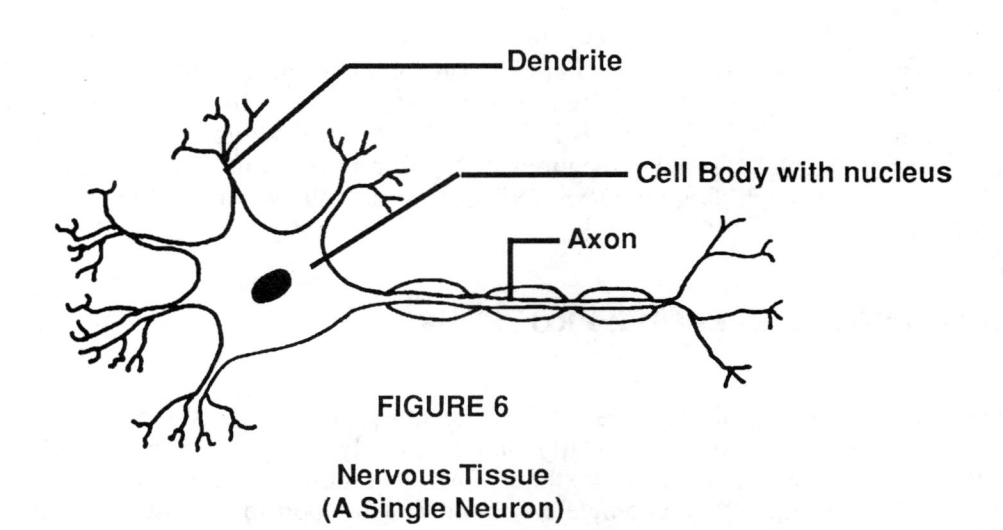

- Haversian System
- Lacunae
- Canaliculi
- Lamella
- Haversian Canal

FIGURE 6

Nervous Tissue
(A Single Neuron)

- Dendrite
- Cell Body with nucleus
- Axon

Animal Tissues 165

canals. The layers of ossified matrix are termed **lamellae** and are composed of hardened calcium salts.

Blood is also classified as a connective tissue because it is the transporting link between all areas of the body. Approximately 55% of the material moving through the circulatory system is liquid **plasma**. The remaining 45% is composed of **erythrocytes** (red blood cells), **leucocytes** (white blood cells), and **platelets**. The red cells do not contain nuclei and are biconcave disc–shaped. The function of erythrocytes is to transport oxygen and carbon dioxide to and from all parts of the body. The white blood cells contain nuclei, are larger than red blood cells, and are divided into those with granules present and those without granules. The function of the leucocytes is to fight infection and inflammation. Those with granules include the **neutrophils**, **basophils** and **eosinophils**. **Lymphocytes** and **monocytes** do not contain granules and are termed agranulocytes. See the textbook for illustrations of blood cells.

Muscle Tissue

The third type of animal tissue is **muscle tissue** that functions to shorten or contract resulting in movement. Three types of muscle tissue are recognized: smooth, skeletal and cardiac. **Smooth muscle** is also referred to as involuntary muscle because it is not under the conscious control of the mind. The cells are elongated and tapered at each end with prominent nuclei visible in stained specimens. Smooth muscle tissue is found in the walls of the stomach, intestines, urinary bladder, uterus, and blood vessels.

Skeletal muscle is referred to as voluntary or striated muscle because it is attached to the bones of the skeleton and it is under the conscious control of the mind. Striated muscle is composed of thousands of elongated, cylinder–shaped cells termed fibers which are composed of thousands of myofibrils. Nuclei are present, and a cell membrane termed the sarcolemma surrounds each fiber. Striations or stripes are present lying at right angles to the long axis of the fibers and myofibrils.

Cardiac muscle is found in the wall of the heart. It is striated in appearance but is involuntary. Each fiber is separated from the next fiber at the end by a thickened section of sarcolemma (muscle cell membrane) termed an intercalated disc. These discs help strengthen the heart muscle which normally contracts about 72 times each minute throughout an individual's life, never stopping to rest.

Nerve Tissue

The fourth type of animal tissue is **nerve tissue**. Nerve tissue is perhaps the least understood of all living tissues. It conducts impulses and relays messages to help in the coordination and regulation of all the body. Figure 6 illustrates a neuron or nerve cell with a large cell body containing the nucleus; small projections called **dendrites** which bring impulses toward the cell body; and a single large extension which carries impulses away from the cell body, the **axon**.

LABORATORY EXERCISE PROCEDURE

Read the instructions in each section of the lab report sheet. Make the drawings as the tissues *appear in the atlas* as neatly and accurately as possible including labels of the indicated structures. Consult the textbook for illustrations of tissues if there is difficulty in locating the structures. When complete, submit the lab report to the instructor for credit.

LABORATORY EXERCISES

I. Space is provided on the report sheet for sketching each type of epithelial tissue. Label the indicated structures and answer the questions on the lab report sheet. Use the *Photo Atlas for Biology* to observe the following epithelial tissues: simple squamous (fig. 122a), stratified squamous (fig. 122b), simple cuboidal (fig. 122a), columnar (fig. 122c) and transitional (fig. 123c). (These figures are found on pages 122–123.)

II. Space is provided on the report sheet to sketch adipose, bone and blood tissue as they appear in the *Atlas*. Label the indicated structure(s) and/or answer the questions. Use the Atlas to observe the following tissues: adipose (fig. 124c), blood (fig. 129d, 130a–f) and bone (fig. 125b). (These figures are found on pages 123, 125, 129, and 130.)

III. Space is provided on the report sheet to sketch a neuron as it appears in fig. 127d. Label the indicated structures and answer the questions.

IV. Space is provided on the report sheet for drawing the three types of muscle tissue as they appear in the *Photo Atlas*. Label the indicated structures and fill in the blanks. Use the *Photo Atlas,* which contains smooth muscle (fig. 127a, b), striated muscle (fig. 126c, d), and cardiac muscle (fig. 126a, b). (These figures are found on pages 126 and 127.)

ANIMAL TISSUES

Report Sheet 1

Name _____

Student ID # _____

Campus _____

Date _____

Note: All sketches **must** reflect the material as it appears in the *Photo Atlas* on pages 122–130. Credit will not be given for drawings from other sources or for drawings on the wrong page or place on the page.

I. In the space provided sketch each type of epithelial tissue. Label the indicated structures and answer the questions.

   A. Simple Squamous Epithelium—Label a nucleus, cell membrane and cytoplasm. Where would you find this tissue in the human body?

   _____

   B. Stratified Squamous Epithelium—Label a nucleus, cell membrane and cytoplasm. Where is this tissue found in the human body?

   _____

168 Exercise 17

Animal Tissues
Report Sheet 2

Student ID # _____

C. Ciliated Epithelium—Label a nucleus, cell membrane, cytoplasm, goblet cell and cilia. Are these cells cuboidal, flat or columnar?

_____

D. Simple Cuboidal Epithelium—Label a nucleus and the cytoplasm. Where would cuboidal epithelium be found in the human body?

_____

E. Transitional Epithelium—Label a nucleus and the cytoplasm. Where would transitional epithelium be found in the human body?

_____

Why? _____

_____

Animal Tissues **169**

Animal Tissues
Report Sheet 3

Student ID # _____

II. Of the connective tissues reviewed, adipose, bone and blood lend themselves to sketching more readily than areolar and cartilage. A space is provided for sketching adipose, bone and blood tissue. Label the indicated structures and/or answer the questions.

   A. Adipose Tissue. Where would adipose tissue be found in the human body? _____

   _____

   List 3 functions of adipose tissue_____

   _____

   B. Bone Tissue—Label an Haversian system, Haversian canal, lacunae, canaliculi and lamella.

Animal Tissues
Report Sheet 4

Student ID # _____

C. Blood—Drawings should be in the appropriate box below with the function filled in.

Red Blood Cells: Function _____	White Blood Cells: Function _____

III. Draw a neuron and label the nucleus, cell body, a dendrite and an axon. What is the function of a dendrite?

_____

An axon? _____

_____

Animal Tissues
Report Sheet 5

Student ID # _____

IV. Space is provided for sketching the three types of muscle tissue. Label the indicated structures on the drawings and fill in the blanks.

    A. Smooth or _____ Muscle—Label a nucleus and cell membrane.

    B. Striated or _____ or _____ Muscle. Label a nucleus, a fiber, sarcolemma and several striations.

    C. Cardiac Muscle—Label a nucleus, striations and some intercalated discs.

18 BIOMECHANICS

LESSON OBJECTIVES

Upon completion of the laboratory exercise the student will be able to:

1. List the functions of the skeletal and muscular systems.

2. Define and/or describe the joint movements shown in the illustration.

3. Given a diagram of the human skeleton, locate the following major bone areas that form the framework of the body: skull, vertebral column, rib cage, pelvis, femur, tibia, fibula, radius, ulna and phalanges.

4. List the three muscle types and briefly describe how you distinguish them from one another.

5. Classify the bones listed in objective 3 as being either a part of the **appendicular** or **axial** skeleton.

6. Answer all questions in the lab report.

MATERIALS NEEDED

A. Blinking
   1. Clock or watch with a second hand
   2. Plastic sheet, such as a piece of Saran Wrap, the size of a sheet of notebook paper
   3. Cotton balls (8–10)
   4. Another person

B. Bone Composition
   1. 5 chicken thigh bones, uncooked
   2. Jars with lids (baby food jars are excellent)
   3. Alcohol (household rubbing alcohol will do)
   4. inegar (white cooking vinegar)
   5. Bleach (such as Clorox)
   6. Water (tap)
   7. Masking tape
   8. Felt tip pen
   9. Forceps*

   *This item is found in your lab kit.

C. Lab manual and textbook

D. *Photo Atlas for Biology*

PREPARATION

Read the discussion which follows carefully before attempting to complete the exercise. Also read the appropriate chapter(s) in the textbook.

DISCUSSION

Skeleton

The skeletal system, which is composed of approximately 206 bones, functions to protect the delicate internal organs of the body, serve as the framework for attachment of muscles, store calcium for future use by the body, and , in some bones, manufacture red blood cells, a process called hemopoiesis. Bones are classified according to their shape such as long bones which include most of the bones of the extremities or limbs, short bones (wrist and ankle), flat bones (cranial bones, sternum, ribs), and irregular bones (vertebrae and some facial bones).

Bone is composed of both inorganic salts such as calcium and phosphorus and organic fibers (proteins). The inorganic salts make bone rigid and hard. The organic fibers make bone tough and somewhat flexible. Approximately 25% of bone is composed of water.

The skeleton is divided into the axial and appendicular divisions. The **axial skeleton** consists of the skull (cranium and face), hyoid bone (bone in the upper neck not attached to any other bone but attached to numerous muscles), vertebral column, sternum, and ribs. The **appendicular skeleton** consists of the shoulder girdle (scapula and clavicle), pelvic girdle (pelvic bones), and upper and lower extremities (limbs).

The skull or cranium consists of 28 bones which make up the face, middle ear, and brain case. These bones are connected by immovable joints or sutures which appear as jagged lines. The **frontal bone** forms the forehead. The **parietals** are two bones joined together at the top of the skull by the **sagittal suture** and located behind the frontal bone. The parietal bones compose most of the top of the head. The **occipital bone** composes the lower back portion of the head. Just below the parietal bones are the **temporal bones**. The temporal bone extends to meet the **zygomatic bone**. The extension of the zygomatic and the temporal forms the zygomatic arch commonly referred to as the cheekbone. **Zygomatic** or **malar bones** form the cheekbones and portions of the **orbit** (the bony cavity or socket in which the eyeball is found). The **ethmoid bone** is a single bone which composes most of the wall separating the nasal cavities. The ethmoid contains openings or air spaces which collectively form the **ethmoidal sinuses**. (This bone is not visible on the drawing in the lab report.) The **maxillae** are two bones which form the upper jaw bone and parts of the orbits, roof of the mouth and walls and floor of the nose. The upper teeth are set into bony sockets in the maxillae. The **mandible** is the largest facial bone and the only movable bone in the skull. It is commonly known as the lower jaw bone. A pair of bones forming the external nose are the **nasal bones**. A pair of small bones located in the front wall of the orbit or eye socket are the **lacrimal bones**. A pair of bones which form the posterior portion of the roof of the mouth are the **palatines**. (These bones are not visible in the drawing of the skull.) A single bone, visible on either side of the skull and anterior to the temporal bones, is the **sphenoid bone**. It forms a large portion of the brain case floor. The **vomer** is a thin, flat bone located along the midline of the nasal cavity. The vomer and the **ethmoid** form the septum which separates the nose into two cavities. The vomer is the more inferior of the two bones.

Twenty-six vertebrae and the intervertebral discs which form strong, partially movable joints make up the vertebral column. There are five divisions of the vertebral column including the cervical, thoracic, lumbar, sacral, and coccyx. Seven **cervical vertebrae** make up the neck region. The first two bones have individual names—the atlas and the axis. The **atlas** supports the skull and is separated from the **axis** by a pivot joint. The **thoracic vertebrae**, the next twelve vertebrae located in the chest region, serve as attachments for the ribs which protect the organs located in the thoracic cavity. The **lumbar vertebrae** are five large vertebrae located in the lower region of the back. This area is sometimes referred to as the small of the back. Five **sacral vertebrae** are fused together to form the triangular shaped sacrum. The **sacrum** is located between the hip bones and is the foundation of the pelvic girdle. Four fused vertebrae compose the **coccyx** or tail bone, a triangular-shaped bone. The vertebral column functions to provide strength and rigidity necessary for support of the head, truck, and upper extremities, provide flexibility, and protects the spinal cord and origins of spinal nerves.

Twelve pairs of ribs plus the sternum form the rib cage. All 24 ribs are attached to the thoracic vertebrae. The first 7 pairs of ribs are also attached to the sternum and are termed true ribs. The next 3 pairs are attached by cartilage to the cartilage of the seventh ribs and are termed false ribs. The last two pairs of ribs are also termed false ribs and floating ribs because they lack any attachment to the sternum.

The arms, hands, and shoulder girdle compose the upper extremities and consist of 32 paired bones. The triangular-shaped shoulder blade is the **scapula**. The collarbone, which articulates with the sternum, is the **clavicle**. The **humerus** is the long bone found in the upper arm. Its rounded head articulates with the scapula in a freely movable ball-and-socket type joint. The **ulna** is the long bone in the lower arm located on the little finger side of the arm. The **radius** is the long bone in the lower arm located on the thumb side of the arm. The **carpals** are eight short bones forming each wrist. The joint between the radius and the carpals is termed ellipsoidal since side-to-side and back-and-forth movement is possible. The five bones forming the hand are the **metacarpals**. The **phalanges** are fourteen bones that form the fingers of each hand—three in each finger, two in each thumb.

Thirty-two pairs of bones forming the legs and pelvic girdle compose the **lower extremities**. You will remember that the sacrum serves as a foundation for the pelvic girdle. Each pelvic or hip bone has a socket, the **acetabulum**, which articulates with the rounded head of the femur in a ball-and-socket type joint. There is a difference in the male and female pelvis. The upper edges of the hip bones, the iliac crests, flare to each side resulting in the wider hips of females. The female sacrum is shorter, broader, and less curved than the male sacrum. These plus other differences result in a wider outlet needed for the birth canal. The thigh bone or **femur** is the longest and heaviest bone in the human body. The smaller of the two lower leg bones is located on the outside of the leg and is called the **fibula**. The larger of the two lower leg bones, the **tibia**, runs parallel to the **fibula**. The tibias or shinbones support most of the weight pressing down on the legs. A small, triangular bone in front of the knee joint forms the kneecap or **patella**. Seven short bones, the **tarsals**, form each ankle and the heel bone or **calcaneus**. Five **metatarsals** form the foot, with the first or innermost bone being larger than the others. It bears more weight than the four smaller metatarsals. Just as in the hand, 14 bones, the **phalanges**, compose the toes in each foot. The big toe has only two heavy phalanges while the other four toes have three phalanges each.

Joints

The joints between two or more bones make up the articular system of the body. Three classes of joints are recognized based on the degree of movement each permits. **Synarthroses** are joints which do not permit movement such as the suture joints of the skull. **Amphiarthroses** are only slightly movable joints such as the articulation between the radius and ulna where the bones are connected by ligaments.

The third class of joints, the **diarthroses**, includes the most familiar joints and are classified according to shape. They are as follows:

1. The **ball–and–socket joint** which consists of a ball–shaped head which fits into a concave socket as in the hip.

2. A **hinge joint** is characterized by flexion and extension motion in a single plane. Examples include the elbow and knee joints.

3. **Pivot joints** allow rotating motion as when the atlas rotates upon the axis resulting in turning of the head.

4. **Condyloid joints** permit movement in two planes at right angles to one another and consist of an oval–shaped condyle which fits into an elliptical–shaped depression as in the wrist.

5. A **saddle joint** is found in the thumb where the articulating surfaces are concave in one direction and convex in the other direction.

6. **Gliding joints**, as exhibited by the joints between carpel bones, permit only gliding movement.

Fractures

Any break in a bone is termed a fracture and several methods of classifying fractures exist among orthopedic surgeons (ortho = straight). The most common types of fractures include:

1. Partial fractures where the break is not completely across a bone.

2. A simple or closed fracture which does not break the skin.

3. Complete fractures in which the bone is broken completely in two.

4. Open or compound fractures in which the bone breaks through the skin.

5. A partial fracture in which the bone bends, termed a greenstick fracture.

Other terms relating to fractures include:

1. Displaced where the alignment of the bone is disturbed.

2. Nondisplaced where the alignment is not disturbed.

3. Comminuted or splintered where small bone fragments are found in the tissues around the break.

4. Oblique, transverse, spiral and longitudinal referring to the direction the fracture takes through the bone.

The healing process following a fracture proceeds slowly and prompt medical attention is desirable to avoid serious complications and promote more rapid repair of the injured areas. The stages which occur during healing of a fracture are presented below in a simplified outline; and, as anyone knows who has ever sustained a fracture, the healing process involves some associated discomfort, inconvenience and long periods of time, directly related to the location and severity of the fracture. These stages include:

1. The formation of a blood clot surrounding and between the ends of the fracture.

2. Formation of a dense fibrous tissue and knitting of the fracture together by a fibrm-cartilaginous mass termed the temporary callus.

3. Replacing the temporary callus by a bony callus.

4. Converting the bony callus into compact bone as healing progresses.

Movements

Movements of the body are a joint action of the muscular and skeletal systems. The type of movement that occurs depends upon the type of joint that is involved. A **gliding movement** occurs when one surface moves back and forth and from side to side over another surface without angular or rotary motion. The ribs and vertebral joints and the joints between the carpals and the tarsal show a gliding movement.

Angular movements increase or decrease the angle between bones. **Flexion** decreases the angle between bones. Examples of flexion include bending the head forward and bending the elbow or knee. **Extension** increases the angle between bones, for example straightening the arm or leg. Special types of extension include **hyperextension**, the continuation of extension beyond the anatomical position, and plantar flexion, extension of the foot at the ankle joint. **Abduction** is the movement of a bone away from the midline of the body. Examples include moving the arm or leg away from the body. **Adduction** is the movement toward the midline of the body with a resulting decrease in the space between the moving parts.

The movement of a bone around its own axis is called **rotation**. No other motion is permitted during rotation. An example of rotation would be shaking the head "no."

The process whereby the distal end of a bone moves in a circle while the proximal end remains stable is **circumduction**. During circumduction the bone describes a cone in the air.

Special movements include the following: (1) **inversion**—the movement of the sole of the foot inward at the ankle joint, (2) **eversion**—the movement of the sole of the foot outward at the ankle joint, (3) **protraction**—the movement of the mandible or clavicle forward on a plane parallel to the ground, (4) **retraction**—the movement of a protracted part of the body backward on a plane parallel to the ground, (5) **supination**—movement of the forearm in which the palm of the hand is turned forward (anterior), (6) **pronation**—moment of the flexed forearm in which the palm is turned backward (posterior), (7) **elevation**—a part of the body moves up, and (8) depression—a part of the body moves downward.

Muscles

Movements of bones occur due to muscle action. Muscles also work to move organs. All muscle, regardless of its type, shows certain common characteristics. These characteristics are irritability, contractility, extensibility, and elasticity. **Irritability** is the ability of a muscle to receive and respond to stimuli. A **stimulus** is a change in the internal or external environment strong enough to initiate a nerve response. **Contractility** is the ability to shorten and thicken (contract) when a sufficient stimulus is received. **Extensibility** is the ability to stretch. **Elasticity** is the ability to return to its original shape after contraction or extension.

The functions of muscle include motion (the change of position of a body part), maintenance of posture, and heat production.

There are three types of muscle—skeletal, visceral, cardiac. **Skeletal muscle**, also called voluntary or striated muscle, is under conscious or voluntary control and shows alternating light and dark lines or striation when viewed under the microscope. **Visceral muscle**, also called smooth or involuntary muscle, is found in the walls of internal organs or viscera and is not under voluntary control. **Cardiac** or heart muscle is a specialized muscle tissue found only in the heart. It is not controlled by the nervous system but has its own built-in contraction mechanism.

LABORATORY EXERCISE PROCEDURE

Carefully complete each exercise recording the required information on the report sheets. When complete, submit the report sheets to the instructor for grading.

LABORATORY EXERCISES

I. Skeleton (Refer to pages 134 and 135 in the *Photo Atlas* for help.)

    A. Label the three figures of the skeleton on the report sheet using the numbers of the terms given on the report sheet. Some terms may be used more than once.

    B. Answer Subsection B and sketch bones in each of the three drawings of arms illustrating the three types of fractures—simple, compound, and comminuted.

II. Blinking Experiment

    A. Objective—Determine if blinking is voluntary, involuntary, or both.

    B. Materials (see page one of this exercise)

    C. Discussion—Muscles attached to the eyelid allow blinking of the eyes. This motion is necessary for protection and lubrication of the eyeball. If the motion is involuntary, there is no conscious control of it. It is voluntary, it can be consciously controlled; that is, you can force your eyelids to close or blink whenever your wish.

    D. Procedure

        1. Watch another person's eyes.

        2. Count and record in Chart 1, in the lab report, the number of times they blink during one minute. Note: Ask the other person not to change the number of times he or she normally blinks.

        3. Repeat this experiment three more times (trials).

        4. Total up the number of blinks, divide by the number of minutes and record the average number occurring in one minute.

        5. Repeat the entire experiment, but this time have the other person watch your eyes and count the number of times you blink.

6. How long can you go without blinking? Have the other person time in seconds how long you can go with blinking. Record your results on this experiment in Chart II. Run four trials and average your results.

7. Change roles again and repeat the experiment with another person.

8. Notice and remember the feeling of your eyes during the above experiment (procedure #6).

9. Hold a piece of Saran Wrap in front of your face and eyes, far enough away from your face to serve as a barrier.

10. Have another person throw a piece of cotton the size of a tennis ball or golf ball at the plastic. Notice and remember if you did or did not blink.

11. Repeat the experiment several times and record the results.

12. Change places with another person and repeat the experiment and record the results.

13. Answer the questions in the lab report.

III. Bone Composition

A. Objective—Explain how the chicken bone experiment shows the importance of both organic and inorganic components of bone.

B. Materials—see page 1 of the exercise

C. Discussion

Calcium is present in all bones. It contributes to the toughness and strength of bones. Certain chemicals are capable of leaching calcium from bone, leaving them soft.

D. Procedure

1. Four different liquids will be tested to see if they soften bone. Before doing the experiment, however, make some guesses about the results. For the liquids listed on the chart in the lab report, mark (on the upper row only) "yes" (will soften) or "no" (will not soften).

2. Check the original hardness of the chicken bone by trying to bend or twist the ends. **NOTE**: Be careful not to snap it. Does the bone bend or twist?

3. Fill a jar with each of the liquids listed. Label each jar with the name of the liquid inside. **CAUTION**: Do not spill bleach on table, skin, or clothing. In case of spills, rinse or flush with water, then add baking soda to area affected.

4. Place a raw bone into each jar. Cover and set aside for three days.

5. After three days, use forceps to remove a bone from one of the liquids.

6. Rinse the bone thoroughly in water and then retest for hardness by twisting and bending.

7. Record in the lower row of the chart if the liquid changed the hardness of the bone. Use "yes" if it did soften, "no" if it did not.

8. Repeat the above steps for each liquid used and record the results of the chart in the lab report.

9. Answer the additional questions on the report sheet.

BIOMECHANICS

Report Sheet 1

Name _____

Student ID # _____

Campus _____

Date _____

I. A. The Skeletal System—These answers will be numbers.

Terms

1. Carpals
2. Cervical vertebra
3. Clavicle
4. Coccyx
5. Femur
6. Fibula
7. Frontal
8. Humerus
9. Lumbar vertebra
10. Mandible
11. Maxilla
12. Metacarpals
13. Metatarsals
14. Nasal
15. Occipital
16. Parietal
17. Patella
18. Pelvis
19. Phalanges
20. Radius
21. Sacrum
22. Scapula
23. Tarsals
24. Temporal
25. Tibia
26. Ulna
27. Vomer
28. Zygomatic
29. Sternum
30. Coccyx
31. Ribs
32. Orbit

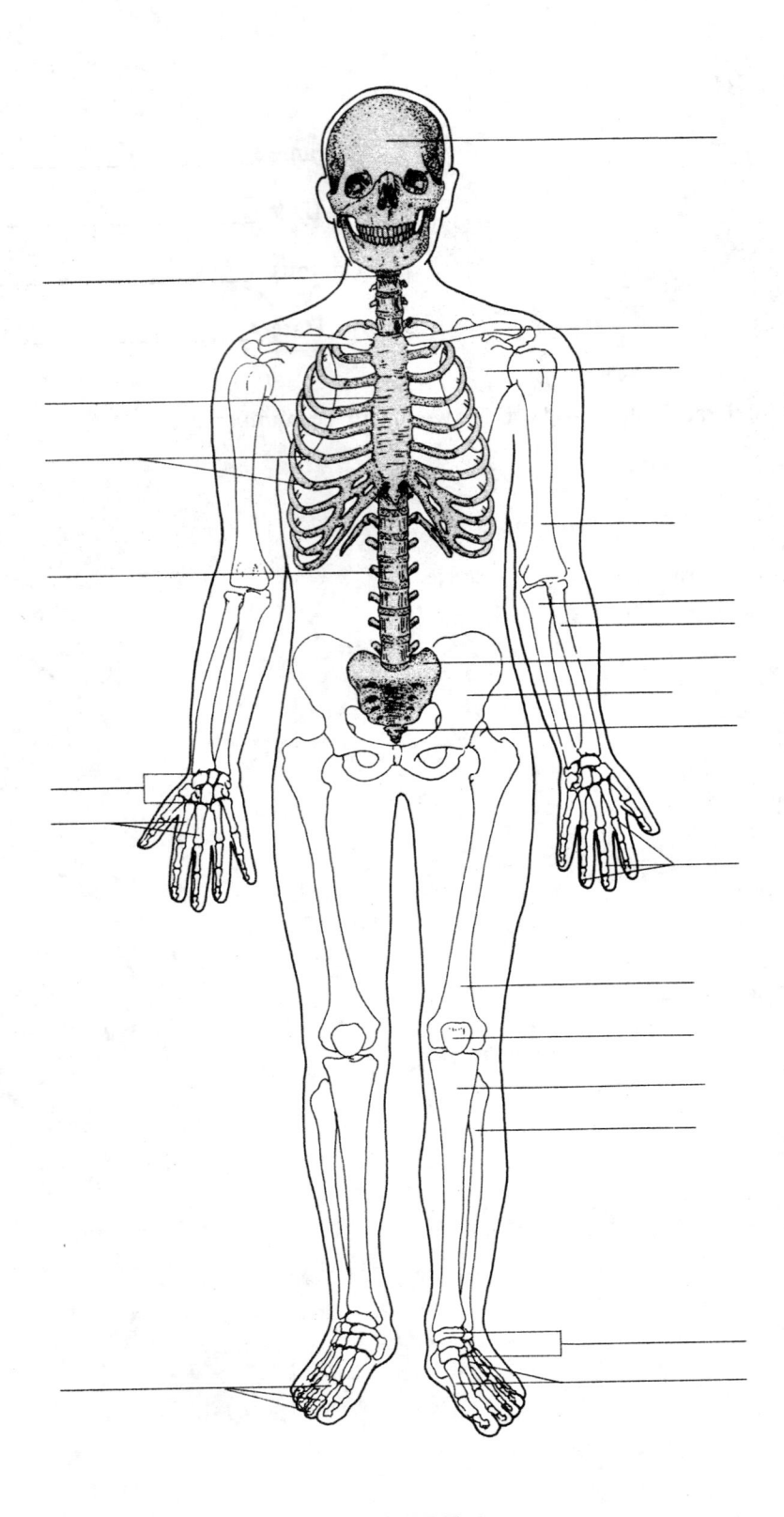

FIGURE 1

Biomechanics
Report Sheet 2

Student ID # _____

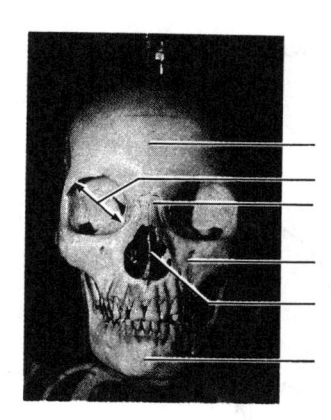

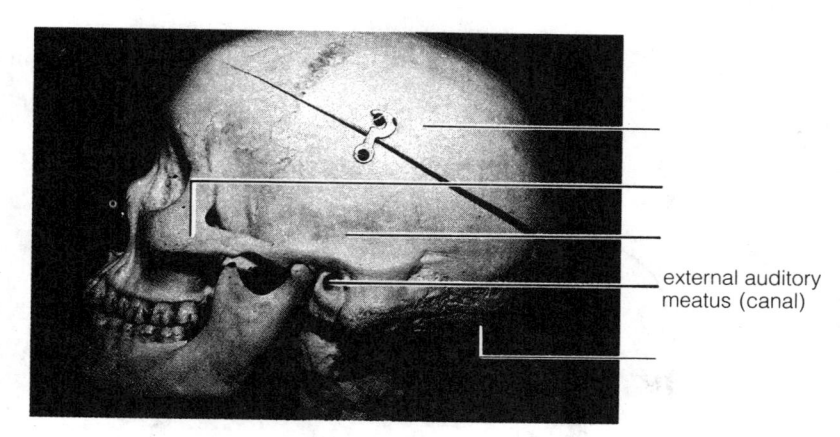

FIGURE 2 (Photos by D. Morton)

external auditory meatus (canal)

I.   B.   Answer the following questions:

1. Describe differences between a female and male pelvis explaining why a difference exists, if one does.

   _____

   _____

2. Designate bones which are part of the axial skeleton with an AX and those which are part of the appendicular skeleton with an APP.

   _____ a. metatarsals   _____ f. nasal

   _____ b. atlas   _____ g. lumbar vertebra

   _____ c. sacrum   _____ h. femur

   _____ d. zygomatic   _____ i. humerus

   _____ e. carpals   _____ j. ulna

3. Identify the type of joint at the following locations:

   _____ a. between the humerus and the radius

   _____ b. at the knee

   _____ c. between the femur and the pelvis

   _____ d. between the occipital and parietal bones

   _____ e. between the clavicle and the humerus

Biomechanics   **183**

Biomechanics
Report Sheet 3

Student ID # _____

4. Sketch the following types of fractures in the upper arm drawings below:

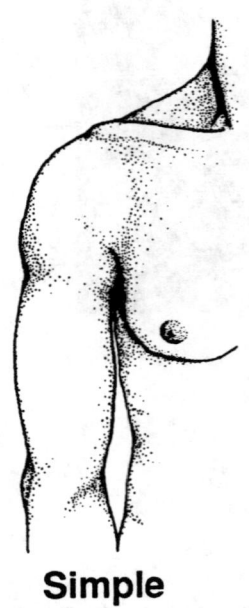

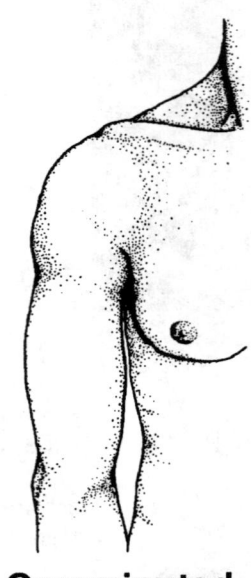

Simple   **Compound**   **Comminuted**

FIGURE 3

II. Blinking

    A. Results

        Chart 1

Number of Blinks
In One Minute

Trial 1	
Trial 2	
Trial 3	
Trial 4	
Total	
Average	

184   Exercise 18

Biomechanics Student ID # _____
Report Sheet 4

Chart II

Number of Blinks
In One Minute

Trial 1	
Trial 2	
Trial 3	
Trial 4	
Total	
Average	

B. Answer the following questions:

1. Is blinking voluntary or involuntary? _____

   Explain: _____
   _____

2. Is blinking protective? _____

   Explain: _____
   _____

3. Did the no blinking experiment show voluntary or involuntary muscle action?

   Explain: _____
   _____

4. How did your eyes feel after not blinking? _____
   _____

5. Do you think that blinking is protective or helpful? _____

   How do you know? _____
   _____

Biomechanics   **185**

Biomechanics
Report Sheet 5

Student ID # _____

6. Did the experiment with the cotton ball and Saran Wrap help prove that blinking is voluntary or involuntary?

   _____

   Explain: _____

   _____

7. What might happen if you were unable to blink? _____

   _____

III. Bone Composition

   A. Results

	Alcohol	Vinegar	Bleach	Water
Guess				
Actual Results				

   B. Conclusions

   1. What liquids removed calcium from the bone? _____

      _____

      Do these liquids have anything in common? _____

   2. A baby's bones are softer than a teenager's. What is the reason for this?

      _____

      _____

   3. Which person is most likely to break a bone when falling—a baby or an old man?

      _____

      Why? _____

      _____

186   Exercise 18

Biomechanics Student ID # _____
Report Sheet 6

IV. Muscles

   A. Classify the following muscles as to type—skeletal, smooth or cardiac.

   _____ a. biceps—muscle of upper arm

   _____ b. artery

   _____ c. stomach

   _____ d. gastrocnemius—calf muscle

   _____ e. intestinal wall muscles

   _____ f. triceps—muscle of upper arm

   _____ g. diaphragm

   _____ h. bladder

   _____ i. heart

   _____ j. lips

   B. Define the following terms:

   a. Flexion _____

   _____

   b. Extension _____

   _____

   c. Abduction _____

   _____

   d. Adduction _____

   _____

   e. Rotation _____

   _____

   f. Circumduction _____

   _____

Biomechanics   187

19 THE CARDIOVASCULAR SYSTEM

LESSON OBJECTIVES

Upon completion of this laboratory, the student will be able to:

1. Describe the structure of the heart.

2. Trace the path of blood through the heart.

3. Distinguish among veins, arteries and capillaries. Take a pulse reading on the wrist or neck and explain the significance of the pulse rate.

4. Describe the sound of the heart.

5. Explain the causes of heart sounds.

6. List 5 factors that influence blood pressure and discuss problems associated with high blood pressure.

7. Examine the components of blood and describe the appearance and function of each.

MATERIALS NEEDED

1. Laboratory Manual
2. Pencil
3. Clock or watch with second hand
4. Rubber stopper—2–hole*
5. Plastic tube *
6. Plastic tube connected to a rubber tube*
7. Yard or meter stick
8. Red food coloring
9. Plastic squeeze bottle*
10. Sink or pan
11. *Photo Atlas for Biology*

   * These items are found in the laboratory kit

PREPARATION

Read the discussion which follows and the appropriate chapter(s) in your textbook.

DISCUSSION

This laboratory exercise examines the cardiovascular system of the human body. This system is designed to transport blood throughout the body. The blood contains gases, nutrients, cells and molecules.

The human circulatory system consists of the heart, blood vessels, blood, lymph, bone marrow, spleen and liver. The major functions of the circulatory system are transportation and protection. It serves as a transporting medium by carrying oxygen (O_2), carbon dioxide (CO_2), nutrients, wastes, water and other materials. It protects the body against disease in two ways. The white blood cells (WBC) fight invading organisms by engulfing them (phagocytosis). Another line of defense is produced by antibodies which are formed by lymphocytes. Some antibodies neutralize toxins produced by foreign organisms (antigens) in our blood. In some instances, antibodies render the antigen susceptible to phagocytosis so that white blood cells can kill them.

Blood

There are approximately 10 to 12 pints of blood in the body depending on the size of the individual. The blood comprises about 5% of the total body weight. A blood donor can donate a pint of blood without ill effects, but the sudden loss of 3 or more pints (as in hemorrhage) can lead to irreversible shock and possible death due to drain damage. Blood is composed of liquid and solid elements. **Plasma**, the liquid part of blood, makes up about 55% of the blood volume. Plasma is 90% water and also contains dissolved foods (glucose, amino acids, fats, minerals and vitamins, dissolved wastes [uric acid and CO_2], and other substances such as proteins (antibodies, clotting proteins, and plasma proteins) and hormones.

The blood contains **erythrocytes** or red blood cells (RBC). There are approximately 700 RBCs for each WBC or every 45 platelets. RBCs give the blood its characteristic red color because they contain hemoglobin. Hemoglobin enables RBCs to carry O_2 and CO_2. The more oxygen the blood contains (oxygenated) the brighter the red color. RBCs are produced in the bone marrow and have a life span of approximately 120 days. The average person has approximately 25 trillion RBCs or about 5,000,000 per cubic mm. of blood. The RBCs also possess specific proteins on their surfaces which can be used in the typing of blood. If the proteins are missing, then the person has type O (the most common type) blood. Mature RBCs have no nuclei.

White blood cells or **leukocytes** fight invading disease–causing organisms by phagocytosis. For this reason the number of WBCs that are present in the blood will vary, depending upon whether or not there is an infection in the body. If there is an infection, the "white count" will be higher than usual. In addition, tissues that contain actively dividing cells, such as the bone marrow, are very susceptible to radiation. The number of WBCs in the blood would be the first sign of over–exposure or radiation damage. WBCs are produced in the bone marrow and have an average life span of 17 days (only 3 or 4 days of this is spend in the blood, some time in tissue fluid and the rest of the time in the lymph fluid). These are the larger of the blood cells. The average person has approximately 8,000 WBCs per cubic mm of blood. The two major groups of WBCs are granulocytes (produced by the bone marrow) and agranulocytes (produced in the lymph tissue).

Platelets or **thrombocytes** are essential to the blood clotting mechanism and are produced in the bone marrow. There are approximately 45 platelets for every 700 RBC. When blood platelets rupture, they cause the formation of thromboplastin, the result of the interaction of substances derived from platelets, blood vessels or tissue cells, and plasma factors, all of which initiates the blood clotting mechanism. Hemophilia is an inheritable (sex–linked) trait whereby the person does not produce sufficient amounts of thromboplastin; the reason being that he (usually) lacks sufficient amount of factor VIII, antihemophilic globulin.

Blood Clotting

Blood clotting results in a clot which is termed a thrombus and consists of a mass of insoluble threads that are organized into a mass which traps other cells. Materials necessary for clot formation include the following: prothrombin and fibrinogen (proteins manufactured by the liver in the presence of vitamin K and then released into plasma), calcium (required from some of the many reactions in the clotting process), 10 or more less well–known clotting factors present in plasma, and thromboplastin (an enzyme which is released from damaged tissue and platelets and stimulates the clotting process). There are three primary stages in clot formation. First, the injured tissues and the platelets at the injury site produce the enzyme **thromboplastin**. Then the thromboplastin serves as the catalyst to convert **prothrombin**, a protein in plasma, into the enzyme **thrombin**. If calcium ions are present, thrombin serves as a catalyst to convert another plasma protein, **fibrinogen**, into insoluble **fibrin** threads. Fibrin, a sticky thread–like substance, traps blood cells and forms a clot.

Blood Types

Different blood types result from differences in specific proteins in the blood plasma and in the red blood cells. The plasma proteins are in the globulin fraction of the blood and are called agglutinins. These substances act upon factors in the red blood cells called agglutinogens, causing them to clump together.

There are thought to be two agglutinins in the plasma which are designated Anti–A and Anti–B. The Anti–A serum will cause any cells having agglutinogen A to clump together, and the Anti–B serum will cause any cells having agglutinogen B to clump together.

Some information relating to the four major blood types is seen in the following chart:

Blood Type	Serum Contains Antibody	Can Give Blood To	Can Receive Blood From
A	B	A, AB	O, A
B	A	B, AB	O, B
AB	none	AB	A, B, AB, O (universal recipient)
O	AB	A, B, AB, O (universal donor)	O

The percent of blood types based on race (in the U.S.A.) is seen in the following chart:

Blood Type	O	A	B	AB
white	45%	41%	10%	4%
black	49%	26%	21%	3%

The Cardiovascular System

The Heart

The heart is a four chambered muscular pump. It beats on an average of 60 to 80 beats per minute and completely circulates the 5 to 6 quarts of blood through the body each 1 1/2 minutes. The heart, being made of muscle tissue, requires a life giving blood supply. The heart gets its supply of blood through the coronary arteries. If a blood clot forms (thrombosis) in these arteries, it results in a heart attack (coronary thrombosis), the severeness of which depends upon how much of the heart muscle is damaged. The heart has a series of one–way valves which are responsible for the movement of the blood through the heart (damage to these valves, as in rheumatic fever, may result in a "heart murmur"). Damage to a heart valve may allow the blood to "back up" which would put a strain on the heart, particularly during vigorous activity.

The "heart sounds" come as a result of the closing of the heart valves, which is related to the muscular contraction (systole) and relaxation (diastole) of the heart.

Blood Vessels

Arteries carry blood from the heart. They have a thick, muscular wall, and blood travels in a "spurting" motion. The largest artery is the aorta. Arteries branch into smaller vessels called arterioles. The most common abnormality of the arteries is "hardening of the arteries," arteriosclerosis, a primary factor in the occurrence of stroke and heart attack. (See the artery illustration.)

Capillaries are the smallest blood vessels. They have thin walls, only one cell thick, which facilitates the exchange of materials. The total surface of our body capillaries approaches 70,000 sq. ft., equivalent to a membrane about 12 miles long and one foot wide. The most common disease is damage due to high blood pressure causing hemorrhage. (See capillary illustration.)

Veins carry blood to the heart. They have thinner walls than do the arteries because they contain less muscle tissue. Veins have one–way valves which prevent blood from flowing back (away from the heart). A common abnormality is varicose veins, also related to hemorrhoids. (See vein illustration.)

Blood Pressure

Blood pressure is the force exerted by blood as it flows through the blood vessels. Factors that influence blood pressure include cardiac output (amount of blood pumped out of the heart in a given period of time), peripheral resistance (resistance of the blood vessels, particularly the capillaries, to the flow of blood through them), viscosity of blood (thickness of blood), volume of blood (amount of blood), and elasticity of arteries (ability of the arteries to recoil during cardiac diastole).

An important part of a physical examination is a determination of the arterial blood pressure. The instrument used for this purpose is a sphygmomanometer. A cuff is tightened around the upper arm by pumping air into it. While listening with a stethoscope to the sounds within the brachial artery, the air in the cuff is slowly released. The blood pressure measure is the pressure at which a pounding sound is first heard—the systolic, and the point at which the pounding sound disappears—the diastolic. The average blood pressure is 120/80 (systolic pressure/diastolic pressure). Systolic pressure is a measure of the force with which blood pushes

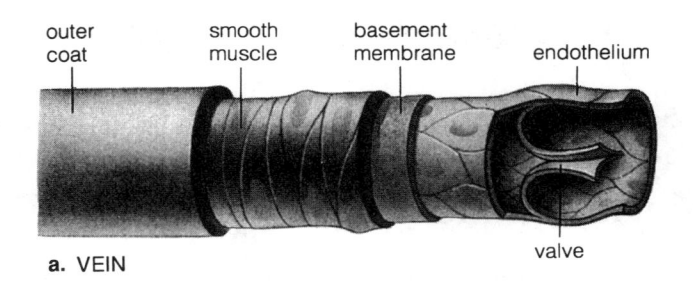

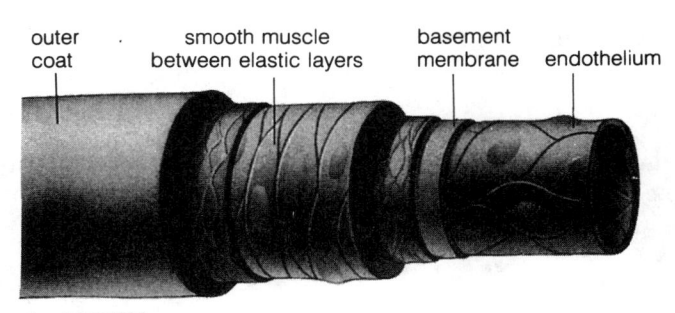

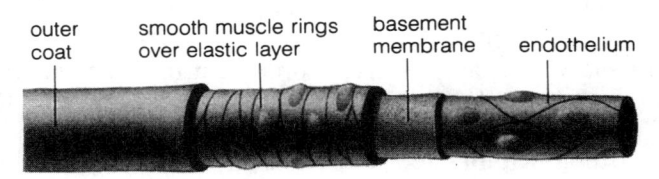

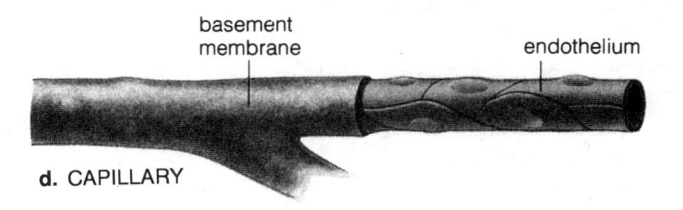

a. VEIN — outer coat, smooth muscle, basement membrane, endothelium, valve

b. ARTERY — outer coat, smooth muscle between elastic layers, basement membrane, endothelium

c. ARTERIOLE — outer coat, smooth muscle rings over elastic layer, basement membrane, endothelium

d. CAPILLARY — basement membrane, endothelium

FIGURE 1

against arterial walls during ventricular contraction. Diastolic pressure measures the force of blood in arteries during ventricular relaxation.

Pulse rate averages between 70 and 90 beats per minute and is the same as the heart rate.

LABORATORY EXERCISE PROCEDURE

Carefully read the directions for each laboratory exercise. Complete each exercise recording the required information on the report sheets. When complete, submit the report sheets to the instructor for grading.

LABORATORY EXERCISES

I. The Heart

   A. Label the drawing of the heart on the report sheet using the terms listed below (some terms may be used twice): (Refer to pp. 119 and 139 in the *Photo Atlas* for help.)

   1. Right atrium
   2. Tricuspid valve
   3. Right ventricle
   4. Ventricular septum
   5. Pulmonary veins
   6. Left atrium
   7. Left ventricle
   8. Bicuspid valve
   9. Aortic semilunar valve
   10. Pulmonary semilunar valve
   11. Right or Left pulmonary artery
   12. Aorta
   13. Superior vena cava
   14. Inferior vena cava

   B. On the report sheet, Section 1B, list in order starting with the inferior and superior ena cava, the path blood follows through the heart and lungs to the aorta.

II. Blood Pressure—Blood within the blood vessels is under pressure. This exercise is designed to demonstrate the difference in the blood pressure of the arteries and veins.

   A. Construction of a *greatly* simplified artificial heart

   1. The materials needed include:
      a. Rubber stopper—2–hole*
      b. Plastic tube attached to a rubber tube*
      c. Plastic tube *
      d. Yard stick or meter stick
      e. Red food coloring
      f. Plastic squeeze bottle
      g. Sink or pan

   * Items found in the laboratory kit.

   2. Procedure for constructing the simplified artificial heart
      a. Fill the plastic squeeze bottle with water.
      b. Add several drops of the red food color to the water in the bottle. Shake gently to mix.

194 Exercise 19

c. Locate the rubber stopper assembly in your laboratory kit. It should resemble the drawing.

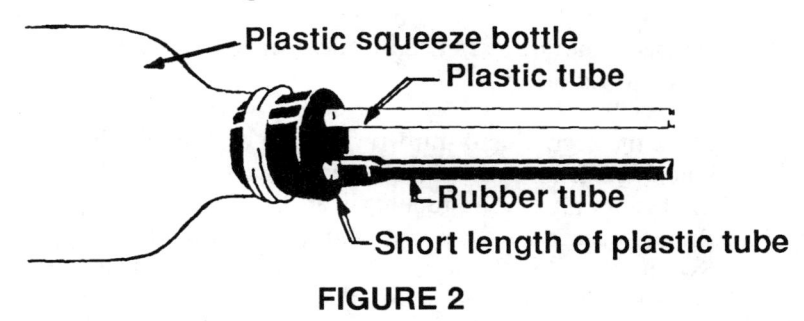

FIGURE 2

d. Insert the rubber stopper assembly into the plastic squeeze bottle opening. The stopper must fit tightly.

3. Experiment Procedure (See Note before proceeding)
   a. Lay the yard stick or meter stick across the sink (or large pan).
   b. Hold the rubber tube with one hand so that it is parallel with the plastic tube and the meter stick.
   c. Holding the bottle in a horizontal position squeeze the bottle. How far does the water travel from the plastic tube? From the rubber tube? Record the results on Section II–A of the lab report sheet.
   d. Repeat the experiment three more times, recording the results each time on the report sheet. Always refill the bottle.
   e. Calculate the average distance that the water traveled for each tube and record this information on the report sheet. (To calculate the average distance add the four distance figures together and divide the sum by 4.)
   f. Answer all the questions on the lab report sheet Section II–B.

Note: Check to be sure that the rubber stopper is secure in the plastic bottle before proceeding. Check all connections to prevent accidents.

III. Pulse—An easy way to count and time a person's heartbeat is by taking the pulse.

   A. Enlist the assistance of a friend or family member in this exercise.

   B. Materials needed

      1. Lab report sheet
      2. Pencil
      3. Clock or watch with a second hand

   C. Procedure—wrist pulse

      1. Locate a large tendon on the inside of the forearm near the wrist
      2. Using the fingers, except the thumb or index finger, gently feel for the pulse just above the point where the large tendon enters the wrist (See Figure 3)
      3. Take the assistant's pulse by counting the number of pulses felt for one full minute. Record this result on the lab report sheet Section III–A.
      4. Repeat the procedure three additional times, record each result on the report sheet and calculate an average. (To calculate an average, add the four pulse figures and divide by 4.)

5. Now, have the assistant take your pulse four times and record the results on the lab report sheet, Section III–A.

D. Neck Pulse—Another area used to determine or "take" the pulse is in the neck using the carotid artery.

1. Place the second and third fingers approximately three inches below the ear lobe. This is on or near the carotid artery. Locate the pulse in this area. Record the results on the lab report sheet Section III–B.
3. Repeat procedure D–2 three more times and average the results.
4. Switch places with the assistant, and ask him or her to repeat the above neck pulse procedure to obtain your pulse.

E. Changes in Pulse Rate

1. Again, enlist the cooperation of a friend or family member.
2. Run in place for one minute.
3. Ask the assistant to take your pulse for one minute as soon as you stop running. Record the pulse rate on the lab report sheet Section III–C.
4. Run in place again for one minute, ask the assistant to take your pulse for one minute, and record the results on the lab report.
5. Repeat procedure 4 six more times, consecutively.
6. Record the results on the lab report sheet after each minute.

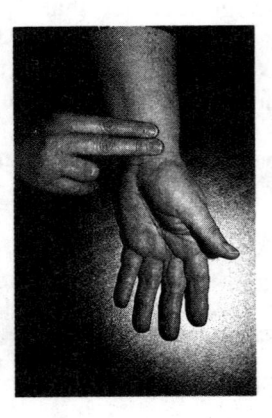

Feeling the radial pulse. (Photo by D. Morton.)

FIGURE 3

IV. Blood

Look at the different views of human blood on pages 124 and 130 of the *Photo Atlas* and answer the appropriate questions on the report sheet.

THE CARDIOVASCULAR SYSTEM

Report Sheet 1

Name _____

Student ID # _____

Campus _____

Date _____

I. The Heart

    A. Label the structures of the heart using the terms listed in Lab procedures Section I–A. These answers will be **numbers.** Use pages 119 and 139 in the *Photo Atlas* for help.

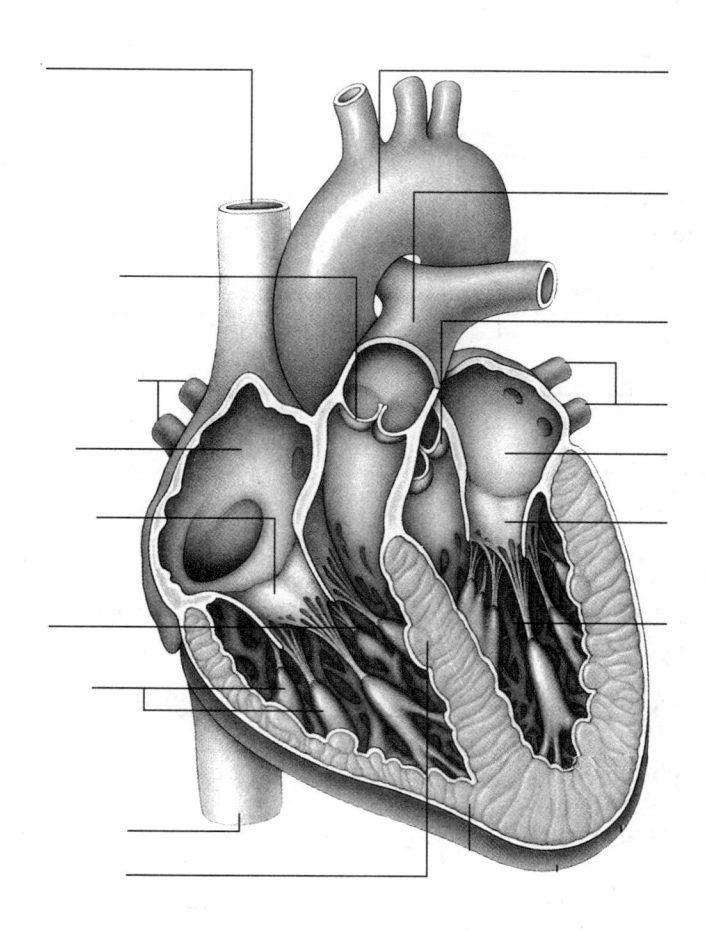

FIGURE 4

The Cardiovascular System   **197**

The Cardiovascular System Student ID # _____
Report Sheet 2

B. List in order the path blood follows through the heart and lungs. Start with—Blood enters the heart via the superior and inferior vena cava which empty into the

1. _____   7. _____

2. _____   8. _____

3. _____   9. _____

4. _____   10. _____

5. _____   11. _____

6. ____lungs_____   12. ____aorta_____

II. Blood Pressure

A. Simplified Artificial Heart Experiment (Results will be recorded in inches or centimeters)

	Result 1	Result 2	Result 3	Result 4	Average
Plastic tube					
Rubber tube					

B. Answer the following questions:

1. Arteries have thick, less flexible, muscular walls than veins which have thinner more flexible walls.

Which of the two tubes corresponds to arteries? _____

Which tube corresponds to veins? _____

2. Material   Human Body Counterpart

Squeeze Bottle   _____

Water   _____

Plastic Tube   _____

Rubber Tube   _____

198   Exercise 19

The Cardiovascular System
Report Sheet 3

Student ID # _____

3. Using the results as a resource, how does the blood pressure in an artery compare to the blood pressure in a vein?

   _____

   _____

4. If the squeeze bottle was not refilled after each trial, would the experimental results be altered?

   _____

   How? _____

   _____

5. What are the five factors that influence (affect) blood pressure?

   a. _____

   b. _____

   c. _____

   d. _____

   e. _____

6. What is the "average" blood pressure reading? _____

7. What is the significance of high blood pressure? _____

   _____

   _____

   _____

III. Pulse

   A.

Wrist Pulse	Result 1	Result 2	Result 3	Result 4	Average
Assistant's Pulse					
Your Pulse					

The Cardiovascular System   199

The Cardiovascular System
Report Sheet 4

Student ID # _____

B.

Neck Pulse	Result 1	Result 2	Result 3	Result 4	Average
Assistant's Pulse					
Your Pulse					

C. Changes in Pulse Rate with Exercise

Your Pulse Each Minute

Minute	1	2	3	4	5	6	7	7

D. Refer to the results in III A & B to answer these questions.

1. Is the wrist pulse exactly the same for each trial? _____

2. Is the neck pulse exactly the same for each trial? _____

3. How does the Average Wrist Pulse compare to the Average Neck Pulse?
   _____

4. Should there be close agreement between the average wrist and neck pulse rate?
   _____

5. Why or why not? _____
   _____

6. Explain the results obtained when the pulse is taken after running?
   _____
   _____
   _____

The Cardiovascular System Student ID # _____
Report Sheet 5

IV. Human Blood Slides—use the *Photo Atlas* and your textbook to answer the following:

    A. Fig. 129d is magnified _____ times. Between the cells in the field of view is a film of plasma which is _____ water plus _____ that assist in immunity, _____ which help maintain blood pressure and _____ which is necessary for clotting of blood. Plasma also contains _____ and minerals which help regulate cell activities and inorganic ions such as _____ , _____ , and _____ . What evidence do you see that RBC's have a biconcave shape? _____

_____

    B. Figs. 130b–130f contain the leucocytes (WBC's). Two of the WBC's have granules in the cytoplasm, the _____ and _____ . The other 3 types of WBC's don't contain granules. In fig. 130d, a monocyte is shown. It is the largest of the WBC's. Look at fig. 130b. The WBC is a neutrophil. These WBCs react _____ (quickly, slowly) to infection and are capable of _____ bacteria. The lymphocyte (fig. 130c) helps fight disease by manufacturing _____ in lymph nodes that fight disease.

    C. Sketch a red blood cell.

The Cardiovascular System   **201**

The Cardiovascular System   Student ID # _____
Report Sheet 6

D. Sketch each of the leukocytes and label them. (You might want to use colored pencils to distinguish among the granular leukocytes.)

E. Platelets are visible in fig. 130a. Platelets are essential to blood clotting and are produced in the _____ _____ . When platelets rupture, the enzyme _____ is produced which converts _____ in plasma into fibrin, a sticky threadlike substance responsible for clot formation.

202   Exercise 19

20 IMMUNITY

LESSON OBJECTIVES

Upon completion of this laboratory exercise the student will be able to:

1. Describe the body's nonspecific defenses against disease.

2. Describe the body's specific defenses against disease.

MATERIALS NEEDED

1. Laboratory Manual
2. *Photo Atlas for Biology*

PREPARATION

Carefully read the discussion which follows as well as the appropriate chapter(s) in your textbook.

DISCUSSION

In order to survive, all animals must have some way to defend themselves against foreign agents, chemicals, and cells. **Nonspecific defense mechanisms** are directed against numerous foreign agents without regard to what those agents are. These mechanisms prevent foreign agents from entering the body and act rapidly to deal with the agents if they do manage to penetrate the body's defenses. **Specific defense mechanisms** are very precise and are specific actions against specific agents. The specific defense mechanisms are collectively called the **immune responses**. Immunity is the study of the body's specific responses to foreign agents.

The nonspecific defense mechanisms involve several processes. The first line of defense against invasion of the body is the skin and mucus membranes. An intact, unbroken skin in the best defense against invasion by chemical agents and especially by pathogens, disease producing organisms. Included in this first line of defense are the skin and its secretions such as sweat and oil and the mucous membranes of the respiratory, digestive, urinary, and reproductive tracts with their numerous secretions.

The second line of defense against foreign agents and pathogens includes the nonspecific internal defenses such as chemicals (interferon and complement), phagocytic cells (white blood cells), natural killer cells, inflammation, and fever. If these defenses are not sufficient, the body responds with the specific immune responses.

The non–specific chemical defenses make it easier for the phagocytic and natural killer (NK) cells to function. **Interferon** increases the resistance of cells to viral attack. **Complement** combines with bacteria to make them more accessible for phagocytosis.

The non–specific cellular defense includes the phagocytic cells and the NK cells. The **phagocytic cells** are white blood cells, including neutrophils and monocytes. These cells are also known as macrophages. These cells also play a crucial role in the immune response by "presenting" parts of the microbe to other cells of the immune system. **Natural killer cells** do not directly attack invading microbes. Instead, NK cells strike at the body's own cells that have been invaded by viruses and destroy those cells. NK cells can also recognize and kill cancerous cells.

The **inflammatory response** is generated against large–scale breaches of the skin or mucous membranes. During this process, damaged cells release histamine into the wounded area. Histamine makes capillary walls leaky and increases blood flow to the injured area. Blood clotting is also initiated to "wall off" the wounded area from the rest of the body. Signs of inflammation include redness (due to increased blood flow to the area), swollen tissues (due to leaking of tissue fluid in the area), heat (due to increased blood flow to the area), and pain. If the wound isn't too large and the microbes reproduce slowly enough, the inflammatory response will keep most of the invaders out of the bloodstream.

Fever has both beneficial effects for the body's defenses and detrimental effects on the invading microbes. Fever also helps fight viral infections.

All of these defenses are nonspecific; that is, their roles are to prevent and/or overcome any and all microbial invasions of the body regardless of the type of microbe. When these defenses aren't sufficient to defend the body, then the immune response is activated.

Immune responses consist of three fundamental steps: (1) recognition of the invader, (2) launching of a successful attack, and (3) retention of the memory of the invader to ward off future infections. The immune response is provided chiefly by two types of lymphocytes (white blood cells) called B cells and T cells. B cells are involved in humoral immunity; T cells, in cell– mediated immunity. B cells produce antibodies which bind to antigens (foreign proteins) and render them susceptible to the T cells, to phagocytic cells, and to the macrophages. T cells include killer T cells which destroy cancer cells and body cells infected by viruses, helper T cells which activate B cells and T cells, suppressor T cells which shut down the immune response, and memory T cells which provide future immunity when the body is exposed to the same antigen. The major function of cell–mediated immunity (T–cell activity) is to destroy the body's own cells when they have become cancerous or have been infected by viruses. B cells include plasma cells which secrete antibodies into the bloodstream and memory B cells which provide future immunity. The immune responses show two distinct characteristics: specificity and memory. The immune response is a specific response to a specific foreign protein. Once the body has activated the immune response in response to a specific antigen, any time the same antigen is presented to the body, an even quicker response will be initiated. This quick response to the second presentation of the antigen is due to the presence of memory T and B cells.

LABORATORY EXERCISE PROCEDURE

Study the *Photo Atlas for Biology* and the appropriate chapters in your textbook. Turn to the report sheet and follow directions for the slides to be investigated.

IMMUNITY

Report Sheet 1

Name _____

Student ID # _____

Campus _____

Date _____

I. Body Defenses Against Infection

   A. What are the three lines of defense against pathogenic organisms?

      1. _____

      2. _____

      3. _____

   B. The Skin—*Photo Atlas*, fig. 128a, pg. 128

      1. What makes the skin a good barrier against disease? _____
      _____

      2. How do you think the arrangement of cells breaking away from the surface of the skin helps keep pathogens out of the body?
      _____
      _____

   C. Epithelial Lining of Respiratory tract—*Photo Atlas*, fig. 122d, pg. 122

      1. What is the function of the goblet cells? _____
      _____

      2. What is the function of the cilia? _____
      _____

   D. Stomach

      1. What secretion(s) produced by the stomach seem to be able to prevent most pathogens from infecting the body?
      _____
      _____

Immunity
Report Sheet 2

Student ID # _____

2. What pathogens seem to be able to survive stomach secretions?

_____

E. Lymph Gland Node—*Photo Atlas*, fig. 131b, pg 131

1. What is the function of the lymph nodes? _____

_____

2. What do you think is going on when you are suffering with swollen nodes?

_____

_____

3. Based on its histology, suggest a function for the appendix (see fig. 131f, pg. 131)

_____

_____

4. Sketch and label a typical lymph node

F. Tonsils

1. What is the common histological feature of tonsilar tissue? _____

_____

2. Where are the tonsils located? _____

_____

206   Exercise 20

Immunity
Report Sheet 3

Student ID # _____

II. *Photo Atlas for Biology*, pp. 129–130. Refer to your textbook for help in answering the following questions.

A. White Blood Cells

1. What is the function of

   a. Eosinophils _____
   _____

   b. Basophils _____
   _____

   c. Monocytes _____
   _____

   d. Polymorphonuclear leukocytes _____
   _____

   e. Lymphocytes _____
   _____

2. Sketch and label each type of white blood cell (Use colored pencils to show the staining of the granules or indicate the color of the granules)

Immunity
Report Sheet 4

Student ID # _____

B. Define phagocytosis _____
_____
_____

C. Immunity

1. Define cell–mediated immunity _____
_____
_____

2. Define humoral immunity _____
_____
_____

3. Describe the function of plasma cells _____
_____
_____

4. Describe the function of T–lymphocytes _____
_____
_____

5. List the types of T–lymphocytes _____
_____
_____

6. What are the two distinctive characteristics of the immune response?
_____
_____

21 PARASITISM

LESSON OBJECTIVES

Upon completion of this laboratory exercise the student will be able to:

1. Classify parasites as ectoparasites or endoparasites and define these two terms.

2. Define mutualism.

3. Describe the parasites which cause malaria, African Sleeping Sickness, tapeworm and hookworm infections, and trichinosis.

4. Explain how the diseases discussed in this lab can be prevented.

5. List stages in the life cycles of the liver fluke and plasmodium.

6. Define intermediate and definitive host.

MATERIALS NEEDED

1. Laboratory Manual
2. *Photo Atlas for Biology*

PREPARATION

Carefully read the discussion which follows. Also read the appropriate chapter(s) in your textbook.

DISCUSSION

Parasites live in or on the living body of a host (an animal or plant) obtaining nourishment from the host. Nutrients can be ingested as solid particles and digested or absorbed in the form of organic molecules through the cell walls from the host's body fluid or tissues. While some parasites cause little or not harm to the host, some produce diseases which destroy the host or produce toxic materials which affect the host's metabolic processes. Parasitic fungi cause most plant diseases. Pathogenic parasites of man and another animals include protozoa, fungi, bacteria and viruses.

Parasites which live on the host's body surface are termed **ectoparasites**. Parasites which live in internal organs or organs ducts leading to the host's body surface (ex: the digestive tract) are termed **endoparasites**. Some parasites are host–specific: these can infect only a single species. Other parasites may require more than one host, an intermediate host which houses the parasite's sexually immature life history stages and a definitive host which houses the sexually mature, reproductive life history stage.

Malaria is caused by **plasmodium**, a protozoan parasite transmitted by the female Anopheles mosquito. An organism which transmits a disease to a host is known as a **vector**. Insects and ticks are examples of vectors. The female Anopheles mosquito feeds on an infected person, one whose blood contains **gametocytes**. Fertilization occurs in the mosquito's gut and a zygote forms a cyst on the exterior of the gut wall. Sporogony, a type of asexual reproduction, occurs in the cyst producing numerous **sporozoites** which move into the mosquito's salivary glands. The mosquito bites a human injecting an anticoagulant and sporozoites into the victim. The anticoagulant prevents coagulation of the victim's blood. In the human body the sporozoites enter the liver, become **cryptozoites** which incubate 10 to 35 days to produce **merozoites**. Merozoites are released into the blood stream and invade the red blood cells, become **trophozoites** which undergo sporogony. The red blood cells rupture releasing toxins into the host's body which bring on the fever and chills characteristic of malaria.

Malaria continues to be a serious disease worldwide but seldom occurs in the United States due to effective methods of destroying the mosquito's breeding places and killing the mosquito. The usual treatment for malaria is chloroquine and rest.

The Chinese liver fluke, *Clonorchis sinensis,* can be found in humans. Heavy infestations may cause serious liver disease and bile duct obstruction. The liver fluke has a complicated life cycle with **adults** residing in the bile ducts of a host for years. Eggs are shed which pass in feces into water. The eggs expel free swimming ciliated **miracidia** which are eaten by snails (the first intermediate host), where they give rise to **cercariae**. The cercariae leave the snail, burrow into fish (the second intermediate host), and encyst as **metacercaria**. When the fish is eaten by man, the metacercaria become **juvenile flukes** which migrate into the bile ducts from the intestines and reach the egg-producing adult stage. Thorough cooking of fish destroys metacercaria.

Tapeworms are classified as flatworms and have ribbon-like bodies composed of an anterior end, the **scolex**, to which is connected many segments termed **proglottids**. The scolex has hooks and suckers by which it attaches itself to the internal wall of the intestine of its host. Each segment or proglottid has male and female reproductive structures. The long chain of proglottids is referred to as a **strobilus**. A strobilus can reach a length of 15 to 30 feet as it hangs in the host's intestine soaking up digested nutrients while the host is deprived of these nutrients. Mature proglottids filled with eggs will be shed in feces to be eaten by hogs, cattle and fish which are the tapeworm's intermediate hosts. The eggs invade the muscles of the intermediate host and form bladder-like structures termed **cysticerci**. When infected muscle (meat) is eaten, the cysticercus everts its scolex which attaches to the mucosa (lining of the intestine) and production of proglottids begins. Infected individuals may have no symptoms or abdominal pain, diarrhea and weight loss may be apparent. Diagnosis is accomplished by finding eggs and/or proglottids in stool samples. Prevention requires thorough cooking of fish, beef and pork. *Taenia saginata* is the beef tapeworm; *Taenia solium* is the pork tapeworm; and *Dipyllothrium latum* is the fish tapeworm.

Necator americanus is the dominant hookworm species in humans in most of the world. Hookworms mature and mate in the small intestine of the host and embryos are expelled with feces. Larvae live in feces and proceed to grow to the third-stage living in the soil. Infection occurs when third-stage larvae contact the skin and burrow into it. If larvae enter blood vessels they are carried to the heart and then to the lungs where they break into the alveoli and bronchi. Ciliary action carries the larvae to the throat and they are swallowed with mucus. Once in the small intestine, larvae attach to the mucosa, grow and undergo two molting stages before becoming sexually mature worms. Approximately 5 weeks is required from infection to egg production.

In the small intestine, the juvenile worm attaches to the mucosa and feeds on blood. The worms move from place to place and it is estimated that about 0.5 cc of blood is consumed by a worm during each feeding. In heavy infestations up to 200 ml of blood may be lost. Obviously iron deficiency anemia will develop if the infestation is severe enough and the host's iron intake is not sufficient to overcome the loss. Common symptoms of hookworm infestation include loss of normal appetite, slight intermittent abdominal pain and a desire to eat soil (geophagy). In very heavy infections, severe protein deficiency, dry skin and hair, edema and "pot–belly" in children with delayed puberty, mental dullness, heart failure, and even death are symptoms.

Hookworm infections are most prevalent in warm climates where residents are poor and unsanitary conditions exist. Where human feces is used as a fertilizer, infection rates run high. Race is another important epidemiological factor with whites being about ten times more susceptible than blacks to hookworm.

Trichinella spirilla is a roundworm which causes trichinosis, a disease contracted by eating improperly cooked pork and/or pork products such as sausage. Improperly cooked pork contains **cysts** filled with larvae which are liberated during digestion and penetrate the mucosa of the small intestine. The larvae mature rapidly and mate. The females then burrow deep into the gut and expel larvae directly into the lymphatic system and blood vessels. The larvae are carried to skeletal muscles, grow to about 1 mm. in length, coil up, encyst, and calcify. Abdominal cramping can be caused by adult worms; but it is the migrating larvae which cause the serious aspects of trichinosis. These symptoms include sore muscles, pain, fever, sweating, chills and prostration. Diagnosis involves blood tests which reveal eosinophils in elevated numbers and the symptoms are treated. In addition to thoroughly cooking pork, freezing it to $-15°$ C for 3 weeks or to $-18°$C for one day will kill encysted larvae.

Trypanosoma brucei gambiense is a protozoan parasite which has as its vector the tsetse fly. The tsetse fly bites an infected animal and then bites a human transferring the trypanosomes to the victim. In humans, trypanosomes live in the blood, lymph nodes, spleen, cerebrospinal fluid and in connective tissues. *Trypanosoma brucei gambiense* invades the central nervous system initiating the chronic sleeping–sickness stage of infection. Symptoms include apathy, mental dullness, disinclination to work with coordination difficulties. The tongue, hands and trunk may exhibit tremors with paralysis or convulsions following. The urge to sleep increases thus giving the disease its name—sleeping sickness. The end stages of the disease are coma and death.

Diagnosis is finding the parasite in the cerebrospinal fluid, blood or bone marrow. Drug therapy is available and satisfactory in most early cases. If the nervous system has become involved, prognosis is poor. Eradication of the tsetse fly is most successful as a control if neighboring countries will cooperate.

Ticks are ectoparasites which feed on the blood of man, domestic animals and pets and belong to the same group as mites, spiders and scorpions—the **arachnids**. Ticks are vectors for *Rickettsia richettsie* which causes **Rocky Mountain Spotted Fever**. Another tick borne rickettsial disease is **lyme disease** carried by the deer tick. Symptoms of Rocky Mountain Spotted Fever include muscular pain, chills, fever, severe headache and prostration. A rash is noted on the wrists, ankles, soles of the feet, palms, and forearms, followed by a generalized body rash. Damage to blood vessels ultimately causes brain and heart damage. Treatment includes antibiotic therapy early on to reduce the mortality rate. Symptoms of lyme disease mimic the symptoms of the flu, including fever, malaise, chills, and muscular pain. A bulls–eye rash at the tick bite site may or may not be present. Untreated, the disease results in rheumatoid symptoms and may even result in a type of rheumatoid arthritis. Early diagnosis and antibiotic treatment is essential to prevent the long term effects of this disease.

Ticks can be removed by applying camphor or kerosene to the tick. When removing ticks care should be exercised so that mouth parts are not left attached to or in the skin. When removing ticks, do not crush the tick and always wash the hands thoroughly after handling ticks.

The flea, an exoparasite, has sharp, piercing mouth parts with which it penetrates the skin of an ape, a bird, cat, dog, man or rat and extracts body fluids. Itching results followed by scratching by the host. The rat flea may carry the bubonic plague from rat to man. Bubonic plague symptoms include chills, fever, headache and body pains. The lymph nodes swell especially in the groin and armpits. These swellings are called buboes from which the disease gets its name. Treatment consists of antibiotic therapy. Rat control and sanitation can help control plague.

The first record of plague in Europe cites an epidemic in Athens in 430 B.C.. In A.D. 262, one of the worst outbreaks occurred in Rome during which 5,000 persons died each day. More than 150,000 persons died of plague in London during 1603 and 1665. Plague entered the United States from South America in 1899. Five deaths occurred in 1980 in the U.S., out of 18 reported cases.

Wheat rust (*Puccini graminis* var. *tritici*), a fungus, is found in over 300 forms of physiologic races. It has an extremely complicated life cycle which involves an alternate host, the common barberry (*Berberis vulgaris*), a shrub. On the wheat plant, the spores of the fungi germinate sending germ tubes into the stem and leaf tissues. The rust utilizes large quantities of food that should be stored in the seed of the wheat. Rusted plants have weakened stems and roots and fall down easily.

In 1950, wheat rust occurred in 16 states in the U.S. attacking nearly all the commercial varieties of wheat grown in North America. New rust-resistant varieties have been developed and eradication of the barberry shrub to eliminate this alternate host has been attempted.

Saprolegnia is a water mold which can live saprophytically in water on dead animal tissue, such as those of fish and insects. It also invades the tissues of living fish when it encounters a wound or abraded skin. The infected fish ultimately dies but the *Saprolegnia* will produce many, many spores which find a new host.

Other fungi, *Epidermophyton* or *Trichophyton* cause *Tinea pedis* or athlete's foot in humans. These are parasites on man and other animals, but may live saprophytically on nonliving organic matter. Athlete's foot can be painful and it should receive prompt treatment. Public pools and gymnasiums should be disinfected daily.

An interesting case of mutualism is found in the case of the relationship of a termite and *Trichonympha,* a complex protozoan which lives in the gut of the termite. Termites eat wood; however, wood contains very small amounts of digestible sugar and protein. Most animals, including the termite, lack the necessary enzymes for digesting the chief constituent of wood – cellulose. The *Trichonympha* can ingest tiny bits of wood and transform them into soluble carbohydrates, some of which can be utilized by the host termite.

LABORATORY EXERCISE PROCEDURE

Turn to the report sheet and follow directions for the figures to be investigated from the *Photo Atlas for Biology*.

PARASITISM

Report Sheet 1

Name _____

Student ID # _____

Campus _____

Date _____

I. Observe *Saprolegnia*. Sketch the mold as it appears in fig. 23b, pg. 23 of the *Photo Atlas for Biology*.

What does a parasite lose by killing its host? _____
_____

II. The scientific name of athlete's foot is _____

What type of organism causes athlete's foot? _____.

III. Examine wheat rust (pg. 30, *Photo Atlas*). What evidence is seen that indicated a broken blister? (Refer to fig. 30a, pg. 30).
_____

Is this specimen capable of infecting other wheat plants if it were in the wheat field?
_____

IV. Observe fig. 21c, pg. 21, of the *Photo Atlas*. This is a relationship of mutual benefit termed
_____

Explain how this is a beneficial arrangement for the termite and *Trichoonympha*.
_____
_____
_____

Parasitism   213

Parasitism
Report Sheet 2

Student ID # _____

V. Why is the flea an ectoparasite? _____

_____

Why are trypanosomes endoparasites? _____

_____

What animal carries a flea that can infect man with plague? _____ Symptoms of bubonic plague include

_____

_____

Treatment consists of _____

How can the rat flea be controlled? _____

_____

Has Bubonic plague been eliminated from the Earth? _____

VI. Refer to pg. 92 of the *Photo Atlas*. Sketch and label a liver fluke. (fig. 92a)

214 Exercise 21

Parasitism
Report Sheet 3

Student ID # _____

The liver fluke can cause _____ and _____ in humans.
List the stages in the complicated life cycle of the fluke starting with

1. adults residing in bile ducts

2.

3.   (eaten by snails)

4.

5.

6.   (in man)

The first intermediate host of the liver fluke is a _____. The second intermediate host is a
_____

VII. Study pg. 96, fig. 96a of the *Photo Atlas*—the *Trichina* worms in muscle. This is a

_____ which causes _____

This disease is contracted by _____
_____

Symptoms of this disease include _____
_____

What are two methods of killing the encysted larvae? _____
_____

VIII. The dominant hookworm species in humans is _____

How is the human infected by a hookworm (how does it enter the human body)? _____
_____

Where does the juvenile hookworm attach in the human? _____

On what does it feed? _____

Parasitism   215

Parasitism
Report Sheet 4

Student ID # _____

What deficiency disease can develop in severe infestations? _____
_____

List three symptoms of the disease. _____
_____

Where are hookworm infections more prevalent? _____
_____

Sketch a hookworm larvae and an adult below.

IX. Malaria is caused by _____ which is transmitted by the

_____. These parasites reach the salivary glands of the

mosquito and are injected into the next victim bitten by the mosquito which is known

as the _____. Why does the mosquito inject saliva when it

bites? _____
_____

216   Exercise 21

Parasitism
Report Sheet 5

Student ID # _____

List the stages in the life cycle of plasmodium starting with

1. gametocytes enter the mosquito as it feeds on an infected person

2.

3.

4.   (in the human liver)

5.

6.   (in red blood cells)

7. toxins released as RBCs burst

Individuals with sickle–cell anemia have a resistance to malaria as the organisms cannot live and reproduce well in the sickle–cells.

X. Examine the tapeworm scolex in fig. 93b, pg. 93 of the *Photo Atlas*. Sketch it, labeling the hooks and suckers. Add the body to the scolex using the discussion as a guide and fig. 93a. Label a proglottid and the strobilus.

A mature proglottid contains _____. Tapeworms attach to the

_____ of its host. The

tapeworm obtains nourishment by _____

_____.

The proper name of the beef tapeworm is: _____.

Parasitism   217

Parasitism
Report Sheet 6

Student ID # _____

The proper name of the pork tapeworm is _____.

The proper name of the fish tapeworm is _____.

22 THE RESPIRATORY SYSTEM

LESSON OBJECTIVES

Upon completion of this laboratory exercise the student will be able to:

1. Identify and explain the function of the following parts of the respiratory system: nose, pharynx, epiglottis, glottis, vocal folds (cords), larynx, trachea, tracheal rings (cartilages), bronchi, bronchioles, alveoli, lungs, diaphragm.

2. Describe the processes of internal and external respiration.

3. Describe the controls over respiration.

4. Measure lung capacity.

5. Determine respiratory rates.

6. Determine changes in the respiratory system due to the effects of smoking.

7. Answer all questions in the lab report.

MATERIALS NEEDED

A. Effect of CO_2 on the respiratory center.
   1. Paper bag
   2. Drinking straws
   3. Clear glass container
   4. Thermometer

B. Lung capacity
   1. Round balloon
   2. Metric ruler

C. Content of exhaled air
   1. Mirror
   2. Lime water*
   3. Drinking straw
   4. Clear glass container

D. Effects of smoking on respiration
   1. Photo Atlas for Biology
   2. Color plates*

E. Respiratory Structures
   1. Photo Atlas for Biology

   * These items are found in the lab kit. (Lime water is made from calcium oxide and water—see Laboratory Exercise Procedure V)

PREPARATION

Carefully read the discussion that follows before attempting to complete the exercise. Also read the appropriate chapter(s) in the textbook.

DISCUSSION

The respiratory system functions in the exchange of gases between the organism and its environment. Oxygen (O_2), required for cellular respiration is taken into the body and carbon dioxide (CO_2), a waste product, is expelled. The basic reaction of cellular respiration is given by the following equation:

$$C_6H_{12}O_6 + 6O_2 \xrightarrow{\text{enzymes}} 6H_2O + CO_2 + ATP \text{ (energy)}$$

(glucose)   (oxygen)   (water)   (carbon dioxide)

The purpose of cellular respiration is to provide energy for cell activities.

The organs of the respiratory system include the nose, pharynx, larynx, trachea, bronchi, bronchioles, and alveoli. The bronchioles and alveoli are located in the lungs.

The **nose** serves as a passageway for air going to and from the lungs. It also serves to filter or clean the air passing through it. Due to the presence of mucus producing cells in the lining of the nose, the nose also adds moisture and warmth to the air on its way to the lungs. Special functions of the nose include aiding in sound formation and containing the receptors giving the sense of smell.

The **pharynx** serves as a passageway for two systems. It transports food from the mouth to the esophagus as a part of the digestive system. It transports gases from the nose to the larynx and back as a part of the respiratory system.

The **larynx** or voice–box is divided into two parts, the epiglottis and the vocal fold or cords. The **epiglottis** prevents food from entering the lungs. The **vocal folds** function in the production of sound. As air is exhaled, the force of air leaving the lungs causes the vocal folds to vibrate resulting in sound. The placement of the teeth and tongue as the air continues out of the body results in speech.

The **trachea** is a hollow tube which connects the larynx and bronchi. Its walls are strengthened and held open constantly by C–shaped rings of cartilage stacked one on top of the other, with the open part of the C facing the esophagus to facilitate swallowing and the peristaltic action of the esophagus. The inner surface of the trachea is lined with a ciliated mucous membrane to keep the internal surface of the air passages free from particulate matter.

The **bronchi** resemble the trachea in structure and are divided into the left and right bronchus which extend into the lungs. Each bronchus branches into smaller tubes called **bronchioles**. The bronchioles also resemble the trachea in structure. They subdivide into several alveolar ducts. The functional unit of the lung and of the respiratory system is the **alveolus** (plural—alveoli) which are the air cells or sacs on the surface of the atrium. Gas exchange occurs between the alveoli and the lung capillaries.

The **lungs** are vascular air sacs which almost completely fill the thoracic cavity. They contain the alveoli and numerous blood vessels. The lungs are covered by a membrane called the **pleura**, which functions to allow movement of the lungs with each breath.

The respiratory system has several functions. First and primarily its role is to supply the cells of the body with oxygen and eliminate excess carbon dioxide. Additional functions include helping to maintain normal blood hydrogen ion (H^+) concentration (pH), helping to maintain normal body temperature, and eliminating minor amounts of water from the body.

The movement of air into and out of the lungs is called ventilation or breathing. It occurs due to pressure differences between the atmospheric air and the air in the lungs. The pressure differences primarily are a result of the contraction and then relaxation of the diaphragm, a sheet-like muscle that separates the chest cavity from the abdominal cavity. Once air is in the lungs, **external respiration** occurs. External respiration involves the passage of oxygen from the alveoli of the lungs to the blood in the lung capillaries and the passage of carbon dioxide from the blood in the lung capillaries to the alveoli of the lungs. This gaseous exchange occurs because of pressure differences. The carbon dioxide is exhaled with the next breath and the oxygen is carried by the blood to the tissues of the body. Loss of carbon dioxide from the body is a significant factor in maintaining normal pH of the blood fluids. **Internal respiration** occurs at the tissue level. Internal respiration involves the passage of oxygen from blood in the body (systemic) capillaries to tissue cells and the passage of carbon dioxide from the tissue cells to the blood in the systemic capillaries. This exchange also occurs due to pressure differences.

The control center for respiration is located in the brain. Control of respiration is both automatic or involuntary and voluntary. An individual can choose to control his respiratory rate or the control can be strictly involuntary. Ultimately, control of respiration is determined by anoxia (marked deficiency or absence of oxygen reaching the cells), acidity of blood, and increased temperature and blood pressure.

Disorders affecting the lungs include lung cancer, emphysema, chronic obstructive pulmonary disease (COPD), bronchitis, and pneumonia, among others. A major cause of lung disfunction is the effect of smoke on lung tissue. Since 1964, the Surgeon General of the United States has issued reports that implicate smoking as a major factor in lung cancer, COPD, emphysema, etc. Smoke effects on lung tissue and respiratory linings include thickening of the epithelium (basal cell hyperplasia), loss of cilia, development of abnormal cell types (which may result in lung cancer), disordered arrangement of cells, and changes in bronchial glands. Lung tissue from smokers will show carbon deposits upon histological examination. The carbon deposits will give the lung tissue a mottled appearance. Emphysema is a lung disorder that is commonly seen in smokers. With emphysema, the lung alveoli are somewhat broken down resulting in an inability to expand upon inhalation or to constrict upon exhalation. As a result the individual is constantly short of breath. It is up to the individual to decide whether or not to smoke; however, based on the effects of smoke on respiratory tissue, the decision should be easy to make.

LABORATORY EXERCISE PROCEDURE

Carefully complete each exercise, recording the required information on the lab report sheets. When complete submit the report sheets to the instructor for grading.

LABORATORY EXERCISES

I. Respiratory System Structures

   A. Identify and give the function of the structures in the respiratory system drawing using the following labels: alveoli, bronchi, diaphragm, larynx, lung, pharynx, sinus, trachea.

   B. Use the *Photo Atlas* and your textbook to answer the questions in the lab report.

II. Effect of Carbon Dioxide on the Respiratory Center

   A. Objective: To demonstrate the effect of changes in carbon dioxide concentration onthe respiratory center.

   B. Materials

      1. Paper or plastic bag
      2. Straws
      3. Glass or other container for water
      4. Thermometer

   C. Discussion

   Hyperventilation (very rapid shallow breathing) removes much of the carbon dioxide normally present in the blood. The increase in the oxygen content of the arterial blood after hyperventilation is negligible, but the carbon dioxide level may fall from a normal value of 44 mm. Hg., to as low as 15 mm.

   Hyperventilation may result in a feeling of dizziness because of cerebral hypoxemia (oxygen deficiency) (due to either decreased blood pressure, to an increase in pH [alkalinity], or the constriction of cerebral vessels caused by the diminished carbon dioxide of the blood).

   D. Procedure

      1. Breathe quietly and normally for three minutes and then note how long the breath can be held after a quiet inspiration. Record on the report sheet.
      2. Place a straw in your mouth, with the other end in a glass container of water. Hold your breath after a quiet inspiration as long as possible. Just before it is necessary to take a breath, begin to sip water. Continue timing, and determine the total time elapsed after the original inspiration until it is necessary to take a breath. Record the results on the report sheet.
      3. Hyperventilate 20 times (breathe shallowly and rapidly 20 times), and then determine how long the breath can be held. Do not continue the experiment if a pronounced feeling of dizziness results. Record on the report sheet.
      4. Place a large paper bag over the mouth and nose. Hyperventilate into the bag 20 times. Record results on the report sheet. **CAUTION: Do not hyperventilate until you are dizzy.** Dizziness can cause you to fall and hurt yourself.
      5. Rapidly run around the building (or in place for several minutes) rapidly. Determine how long the breath can be held immediately after the cessation of this exercise. Record the results on the report sheet.

6. Enter all times in the table under Results and Questions on the report sheet. In the discussion column on the Report Sheet explain the results obtained in this experiment. Use the fact that the respiratory center is very sensitive to slight changes in the carbon dioxide content of the blood.
7. Record the temperature of the inhaled air (room temperature) on the report sheet.
8. Record the temperature of the exhaled air (exhale on the thermometer several times) on the report sheet.
9. Explain what caused the difference between the temperature of inhaled and exhaled air on the report sheet.

III. Measuring Lung Capacity

   A. Objective: To measure lung capacity

   B. Materials: round balloon and metric ruler

   C. Procedure

      1. Stretch the balloon so that it will be easier to blow up.
      2. After taking a very deep breath, blow up the balloon as fully as possible, but with one single breath. Exhale as much air from the lungs as possible.
      3. While holding the end of the balloon to keep air from escaping, measure the diameter of the balloon.
      4. Record the results on the report sheet.

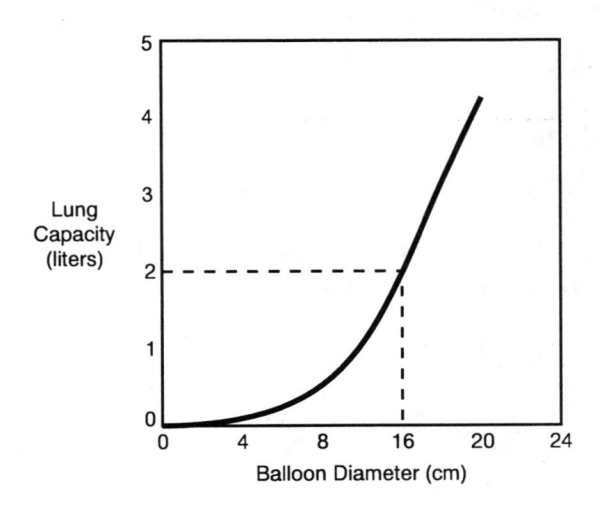

      5. Use the graph to convert the balloon diameter to lung capacity. In the example shown on the graph, a balloon diameter of 16 cm. represents lung capacity of 2 liters.
      6. Repeat the experiment at least five more times and average the results and record them on the report sheet.

IV. Respiration Rate

   A. Objective: To determine the respiration rates of several other people.

   B. Materials—Watch or clock with a second hand

The Respiratory System   223

C. Procdure

1. Observe the respiration rate of a person breathing normally while at rest.
2. Count the number of times the person breathes in one minute.
3. Repeat the experiment three more times with three different people.
4. Record the results. (The average respiration rate at rest is about 15 times per minute.)
5. Count the respiration rate for each person after they have run in place for one minute and record the results.
6. Answer the questions in the lab report.

V. Content of Exhaled Air

A. Objective: To determine the content of exhaled air.

B. Materials

1. A mirror
2. Lime water (dissolve 1/2 teaspoon of calcium oxide in 1/2 cup water)
3. Drinking straw
4. Clear glass container

C. Procedure

1. Take a deep breath through the nose and mouth
2. Exhale against the surface of the mirror
3. Take another deep breath through the nose and mouth
4. Exhale through a straw into lime water
5. Answer the questions in the lab report

VI. Effects of Smoking on the Respiratory System

A. Objective—To determine some of the effects of smoking on the respiratory system

B. Materials—Color plates of emphysema lung and smoker's lung with carbon deposits and *Photo Atlas for Biology*

C. Procedure—Answer the questions on the report sheet.

FIGURE 1

The Respiratory System

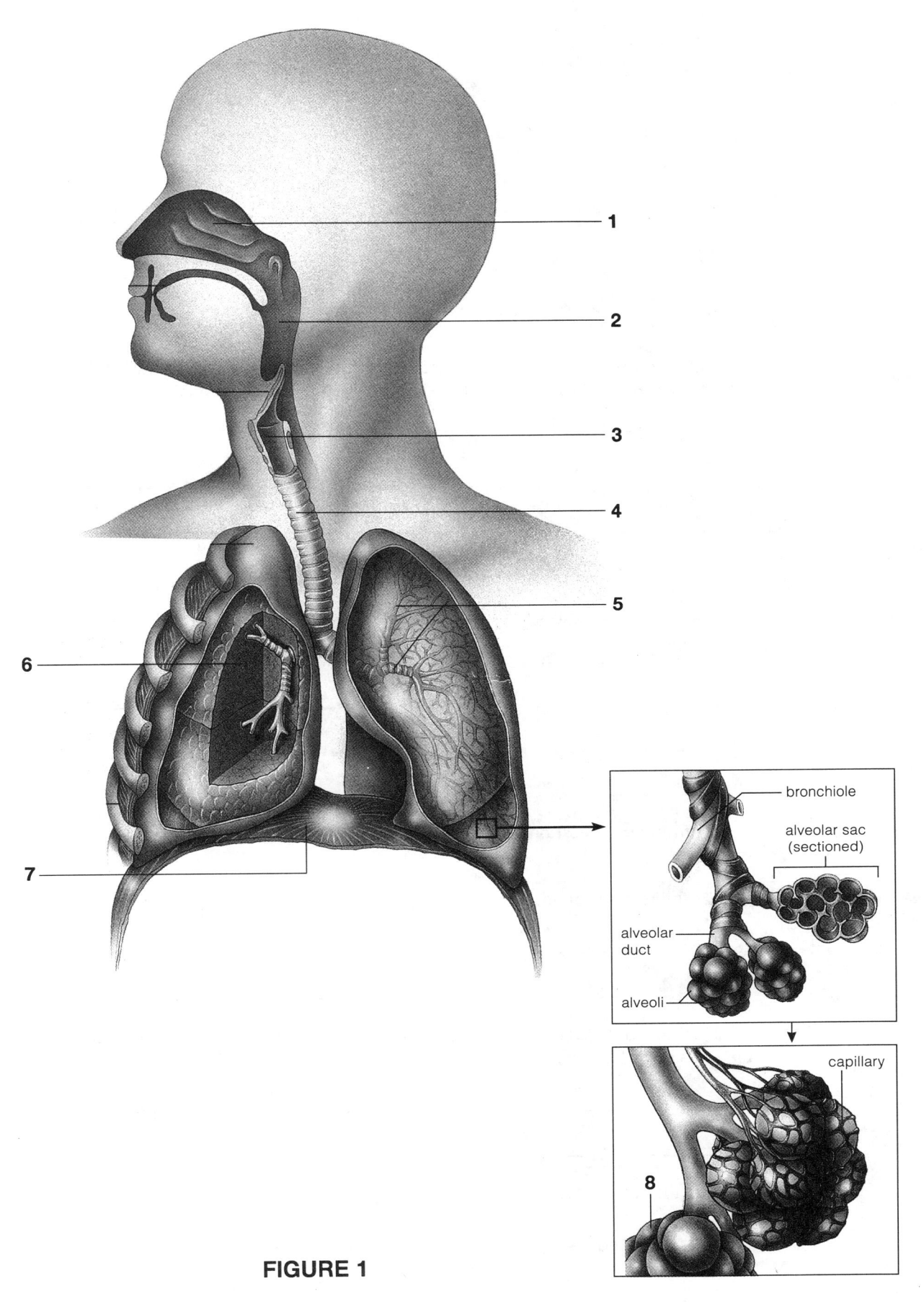

THE RESPIRATORY SYSTEM   Report Sheet 1

Name _____

Student ID # _____

Campus _____

Date _____

I. Respiratory Structures

   A. Identify and give the function of the structures in the respiratory system drawing. Refer to pages 118 and 120 in the *Photo Atlas* for help.

   Structure   Function

   1. _____   _____

   2. _____   _____

   3. _____   _____

   4. _____   _____

   5. _____   _____

   6. _____   _____

   7. _____   _____

   8. _____   _____

   B. Answer the following questions—refer to *Photo Atlas for Biology* and your textbook

   1. Define the term respiration _____

   _____

   _____

   2. What is the advantage to having the cartilage of the trachea C–shaped rather than in a complete ring?

   _____

   _____

   3. What is the function of the goblet cells in the respiratory passages? _____

   _____

The Respiratory System
Report Sheet 2

Student ID # _____

4. Describe the work of the capillary networks _____

   _____

   _____

   _____

5. What is the function of the dust cells in the lungs?

   _____

   _____

   _____

6. What structures within the cell accomplish the slow, quiet energy release needed by cells?

   _____

7. Draw the structure of the mitochondrion and label the site(s) of enzyme location. Use *Photo Atlas for Biology*, fig. 8b, pg. 8 as your guide.

The Respiratory System Student ID # _____
Report Sheet 3

II. Answer the following from Experiment I.

A. Results: Fill in the following as the experiment is performed.

Procedure	Time Breath Held	Explain Reason for Result
Quiet Breathing		This sets up normal values
Sipping Water		
Hyperventilation		
Paper Bag		
Running		
Temperature of	Inhaled Air _____ Exhaled Air _____	

B. Conclusions

_____ 1. The major gas in exhaled air is

a. O_2 b. CO_2 c. H_2O d. Two of these are correct

_____ 2. Holding one's breath causes a buildup of _____ in the blood.

a. O_2 b. CO_2 c. H_2O d. N_2

_____ 3. Hyperventilation causes a buildup of _____ in the blood.

a. O_2 b. CO_2 c. H_2O d. N_2

_____ 4. CO_2 is necessary for normal respiration.

a. true   b. false

The Respiratory System Student ID # _____
Report Sheet 4

III. Answer the following from Experiment II.

   A. Results: Fill in the chart as the experiment is performed.

   Balloon Diameter

Trial 1	
Trial 2	
Trial 3	
Trial 3	
Trial 4	

   Average _____

   B. Answer the following questions:

   1. What is the average lung capacity? _____

   2. Why must the balloon diameter be converted from cm. to liters? _____

   _____

   3. What variation in lung capacity might be expected in a heavy smoker? _____

   _____

   4. What variation in lung capacity might be expected in a seven–day per week jogger?

   _____

   Why?

   _____

   _____

The Respiratory System Student ID # _____
Report Sheet 5

IV. Answer the following from Experiment III

   A. Results

   Respiration Rate

Person	At Rest	After Exercise
1		
2		
3		
4		

   B. Answer the questions:

   1. Do all people have the same respiration rate at rest? _____

      After exercise? _____

   2. What contributes to differences in respiration rate of different individuals?
      _____
      _____

   3. Would a physically fit person tend to have a slower or faster respiration rate than a person who never exercises?
      _____

      Why? _____
      _____

   4. List some factors that might affect a person's respiration rate. _____
      _____
      _____

V. Answer the questions from Experiment IV

   1. Describe what occurred when breathing on the mirror. _____
      _____

The Respiratory System　　　　　　　　Student ID # _____
Report Sheet 6

2. Does the mirror experiment indicate the presence of a particular substance in exhaled air?

   _____ If so, what is the substance? _____

3. Describe what happened to the lime water when you breathed into it. _____

   _____

   _____

4. Does this show that there is a chemical reaction between some substance in the exhaled air and the lime water?

   _____

5. Describe the results? _____

   _____

6. The following chemical reaction indicates the interaction between carbon dioxide and lime:

$$CO_2 + Ca(OH)_2 \longrightarrow CaCO_3 + H_2O$$

   (carbon dioxide)　　(lime)　　　　　　(calcium carbonate)　(water)

   In a water solution, $CaCO_3$ causes the water to appear cloudy. Based upon your observation, does the lime water experiment indicate an exhaled product of air?

   _____ If so, what is the product? _____

VI. Discussion Questions

   1. Describe what is meant by external respiration. _____

      _____

      _____

   2. Describe what is meant by internal respiration. _____

      _____

      _____

The Respiratory System
Report Sheet 7

Student ID # _____

3. Describe the controls over respiration. _____

   _____

   _____

4. Define the following terms:

   a. Anoxia _____

   _____

   _____

   b. Alveolus _____

   _____

   _____

   c. Hypoxemia _____

   _____

   _____

VII. Smoking and Health (Use the *Photo Atlas for Biology*, fig. 122d, pg. 122 and your textbook for help)

   Answer the following questions:

   1. What is pseudo–stratified ciliated columnar epithelium? _____

      _____

      Where can it be found? _____

      _____

      What is the function of the cilia? _____

      _____

      _____

232   Exercise 22

The Respiratory System Student ID # _____
Report Sheet 8

2. What effect does smoke have on cilia? _____

_____

_____

3. What effect does smoke have on bronchial epithelium? _____

_____

_____

4. Why must the walls of the air sacs be thin to accomplish oxygen exchange?

_____

_____

_____

5. Define emphysema. _____

_____

_____

6. What differences can be seen between an emphysema lung and a normal lung?

_____

_____

_____

7. What differences can be seen between a smoker's lung and a normal lung? _____

_____

_____

_____

The Respiratory System Student ID # _____
Report Sheet 9

8. Based upon the histology of the smoker's lung and the emphysema lung, what would you predict about lung functioning?

_____

_____

_____

_____

9. How do you explain the fact that cigarette consumption has begun to decline but the death rate continues to climb?

_____

_____

_____

_____

23 THE DIGESTIVE SYSTEM

LESSON OBJECTIVES

Upon completion of this laboratory exercise, the student will be able to:

1. List the major components of the digestive system, beginning with the mouth and proceeding in order down the G.I. tract.

2. Describe the structure and function of the major components of the digestive system.

3. Define chemical digestion.

4. List or identify the end products of carbohydrate, protein, and lipid digestion.

5. List the primary functions of carbohydrates, fats, and proteins in the human body.

6. Answer questions over the discussion material.

MATERIALS NEEDED

A. Digestive organ structure
   1. *Photo Atlas for Biology*
   2. Color plates*

B. Test for starch exercise
   1. Slice of white bread
   2. Slice of raw potato
   3. Several grains of rice, uncooked
   4. Slice of raw apple
   5. Several oatmeal flakes, uncooked (may use instant)
   6. A soda cracker
   7. A small glass of water
   8. 1/2 tsp. cornstarch
   9. Medicine dropper
   10. 2 glasses (must be able to see through them)
   11. Tincture of iodine*
   12. 5 foodstuffs of your choice

C. Test for sugar
   1. Dextrose (sugar)*
   2. Benedict's solution*
   3. Small sauce pan
   4. Heat resistant jar
   5. Test tube*

*Items are found in the laboratory kit.

D. Conversion of starch to sugar exercise
   1. Water
   2. A soda cracker
   3. A small sauce pan
   4. A heat-resistant jar, small enough to fit in the sauce pan
   5. Test tubes*
   6. Benedict's solution*

E. Test for fat
   1. Sheet of unglazed paper (such as paper towel or a napkin)
   2. Bacon (raw slice)
   3. Few drops of vegetable oil

*Items are found in the laboratory kit.

PREPARATION

Read the discussion carefully before attempting the exercises. Also read the appropriate chapter(s) in the textbook.

DISCUSSION

This laboratory is a study of the human digestive system. The digestive system processes food and prepares it for use in the body. The process of food preparation for use is called **digestion**. Digestion involves several basic steps: ingestion, digestion (mechanical and chemical), propulsion, absorption, and defecation. The first step in digestion is **ingestion** or eating—the taking in of food. Digestion of the ingested food involves both mechanical and chemical processes. **Mechanical digestion** involves the physical breakdown of food into smaller pieces which increases the surface area of the food. **Chemical digestion** is an enzymatic breakdown of food into less chemically complex components. Mechanical digestion includes **mastication** or chewing (occurs in the mouth), and **deglutition** or swallowing (occurs in the pharynx and esophagus). Chemical digestion occurs in the mouth through the action of saliva (digestion of carbohydrates), the stomach (digestion of proteins), and the small intestine (digestion of proteins, carbohydrates, lipids and nucleases). The digestive activity in the small intestine is due to enzymes produced primarily by the pancreas. Some digestive activity occurs in the large intestine due to bacterial action. **Propulsion** is the movement of food stuff through the digestive system. Propulsive actions include **peristalsis**, the rhythmic movement of the muscles of the digestive tract to mix food and transport it throughout the digestive tract, and **churning** (occurring in the stomach, small intestine, and large intestine) which is primarily a mixing action. **Absorption**, the movement of the end-products of digestion into the circulatory and lymphatic systems occurs primarily in the small intestine, although some absorption does occur in the mouth, stomach, and large intestine. The final process is **defecation**, the elimination from the body of indigestible wastes in the form of feces.

The digestive tract (also called gastrointestinal tract or alimentary canal) consists of a 30' tube extending from the mouth to the anus. Major sections of this tube are the mouth, pharynx, esophagus, stomach, small intestine, large intestine, rectum, and anus. Accessory organs to the tract include the teeth, tongue, salivary glands, liver, gall bladder, and pancreas. All of the accessory organs except the teeth and tongue are found outside the G.I. tract and produce or store secretions that aid chemical digestion.

The **mouth** is formed by the cheeks, hard and soft palates and the tongue. The cheeks make up the sides of the mouth with the palate being the roof and the tongue the base of the mouth. The **hard palate** is formed by bones of the skull and is lined or covered by a mucous membrane. The **soft palate** is a muscular partition between the oropharynx and the nasopharynx. It too is lined with a mucous membrane. The **tongue** is skeletal muscle covered by a mucous membrane. The mucous membrane of the tongue contains small projections called **papillae**. These papillae contain the **taste buds** – areas sensitive to salt, sour, sweet, and bitter stimuli. The taste areas are sweet at the tip of the tongue, bitter at the back of the tongue, salt on the sides of the tongue towards the front and sour on the sides of the tongue toward the back.

Closely associated with the mouth are the **salivary glands**. There are three pairs of salivary glands. The **parotids** are located under and in front of the ears. They secrete mostly water and amylase, a carbohydrate digesting enzyme. The **submandibulars** lie beneath the base of the tongue and secrete amylase and mucous (used for lubrication). The **sublinguals** lie beneath the anterior floor of the mouth and secrete mostly mucous. Collectively, the secretions of the salivary glands are called **saliva**. Saliva, which contains the amylase ptyalin, a carbohydrate digesting enzyme, serves to initiate chemical digestion in the mouth. Other accessory organs associated with the mouth are the teeth which serve to mechanically break down food by the process of chewing and the tongue which serves to place food in contact with the teeth and to initiate the swallowing process.

The **pharynx** or throat serves as a passageway for food coming from the mouth and going to the esophagus. The pharynx has three parts—the nasopharynx, the oropharynx, and the laryngopharynx. Each section is named on the basis of its location.

Food, now called a **bolus**, goes from the pharynx into the esophagus. Histologically, the esophagus and the remaining digestive organs show four layers: mucosa, submucosa, muscularis, and adventitia or serosa. The mucosa layer consists of epithelium (type depends upon the organ and its function). The submucosa consists of connective tissue and possibly glands and/or lymphatic nodules. The muscularis consists of two layers of smooth muscle (except in the stomach where there are three layers). The adventitia (in the esophagus) or the serosa (in the remaining organs) is an epithelium that covers the outer surface of the organ.

The **esophagus** is a muscular collapsible mucus–lined tube which connects to the stomach. It serves as a passageway for the bolus on its way to the stomach. Its mucosa consists of stratified squamous epithelium for protection against the undigested bolus. The **stomach** is a J–shaped enlargement of the G.I. tract. It lies directly under the diaphragm and attaches to both the esophagus and to the small intestine. The mucosa of the stomach consists of simple columnar epithelium. When the stomach is empty, the internal lining is arranged into folds of mucus tissue called **rugae**. The rugae serve to increase surface area without increasing the size of the organ. Within the fold of rugae are the gastric glands which secrete pepsinogen and hydrochloric acid (HCl). Together the pepsinogen and HCl form the protease pepsin, an enzyme which serves to chemically break down proteins. Mucous cells in the rugae secrete mucus for lubrication and the intrinsic factor for vitamin B_{12} absorption. The chemical digestion that occurs in the stomach is the breakdown of proteins by the enzyme pepsin. The mechanical digestion is a mixing of the stomach contents by the process of churning and the movement of **chyme**, the name for the food material which has now been partially digested by both an amylase and a protease, to the small intestine by the process of peristalsis.

By peristalsis, chyme enters the small intestine. The small intestine is divided into three sections—the **duodenum**, which is attached to the stomach, the **jejunum** and the **ileum**, which is attached to the large intestine. The remaining chemical breakdown of food occurs in the small intestine, in the duodenum and the jejunum. The lining of the small intestine contains small finger-like projections called **villi**. They serve to increase surface area without increasing size. Located among the villi, particularly of the duodenum, but also in the first portion of the jejunum, are digestive glands that secrete amylases, proteases, and lipases – carbohydrate, protein, and fat digesting enzymes. These enzymes along with the pancreatic juices and bile serve to complete the chemical breakdown of food. Pancreatic juices come from the **pancreas**, a tear-drop shaped organ which lies just posterior to the stomach. The pancreas has two major functions. It serves as an **endocrine gland**, a gland that secretes its chemicals, hormones, directly into the blood stream, and as an **exocrine gland**, a gland that secretes its chemicals via a duct to the target site. As an endocrine gland, the pancreas produces insulin and glucagon, both of which regulate blood sugar level. As an exocrine gland, the pancreas secretes digestive enzymes which are carried via the pancreatic duct to the duodenum to aid in the final chemical breakdown of food (chyme).

The **liver**, the largest organ in the body, is located just under the diaphragm, a muscle sheet that separates the thoracic or chest cavity from the abdominal cavity. The liver serves to produce most blood proteins, to remove worn out red blood cells, to detoxify poisons, to store nutrients such as glycogen and fat, to store copper, iron, and vitamins A, D, E, and K, and to produce bile. **Bile** serves as a digestive chemical. It contains no enzymes but is a salt solution that emulsifies or dissolves fats. Bile is transferred from the liver to the gall bladder for storage and is released through the cystic duct from the gall bladder into the common hepatic duct which empties into the duodenum.

After the chyme from the stomach has been digested in the duodenum, the end-products of digestion are absorbed into the bloodstream or lymphatic system. This absorption occurs in the jejunum and ileum. The end-product of carbohydrate digestion, glucose, and the end-products of protein digestion, amino acids, are absorbed into the bloodstream. The end-products of fat digestion, glycerol and fatty acids, are absorbed into the lymphatic system. Food that could not be digested such as cellulose from uncooked plants, and water remain in the ileum as waste material. By peristalsis, this waste material is carried to the **large intestine**.

The first part of the large intestine is a blind pouch called the **cecum**. Attached to the cecum is the **appendix**, an organ that probably functions in disease prevention due to its large amount of lymphatic tissue but an organ whose function is not really known. The cecum is attached to the ileum and to the ascending colon, the first major part of the large intestine. The ascending colon is followed by the transverse colon, the descending colon, the sigmoid colon, and finally the rectum and anus. The liquid wastes from the ileum pass into the colon by peristalsis and water is removed until the waste becomes semi-solid. Water removal occurs primarily in the ascending colon. The semi-solid waste material travels through the remaining parts of the large intestine to the rectum where it is stored until released from the body via the anus by a process called **defecation**. The indigestible semi-solid waste material is called **feces**.

The digestive system functions to break down carbohydrates, lipids, and proteins into usable forms for the body. Carbohydrates furnish energy for all body activities and functions. The usable form of carbohydrate is glucose. Lipids (fats) serve as a reserve source of energy as well as providing structural molecules for building membranes and other tissues in the body. The usable forms of lipids are glycerol and fatty acids. Proteins build, repair and maintain body tissues; build regulating substances such as hormones and enzymes; and can, when needed, furnish energy for body use. The usable forms of proteins are amino acids. These breakdown products – glucose, glycerol, fatty acids, and amino acids are used by the body to build, repair, and maintain all body activities and parts.

Location	Glands	Secretions	Enzymes	Chemical Activity	Mechanical Activity
Mouth	salivary	saliva	ptyalin	changes starch to disaccharides	chewing swallowing
Pharynx	tonsils (lymph organs)	none	none	none	swallowing
Esophagus	mucous	mucus	none	lubrication	swallowing peristalsis
Stomach	gastric	protease (pepsin)	pepsin	changes proteins to peptides and proteases	churning peristalsis
	gastric	HCl	none	dissolves minerals, kills bacteria	
	mucous	mucus	none	lubrication	
Small Intestine	intestinal	sulcus entericus (intestinal juices)	erepsin (protease)	changes peptides to amino acids	churning in duodenum— peristalsis throughout
			maltose (amylase)	changes maltose to glucose	
	liver	bile	none	emulsifies fat	
	pancreas	pancreatic juices	trypsin (protease	forms peptides (smallest proteins)	
			amylase	forms maltose	
			lipase	changes fat droplets to glycerol and fatty acids	
	mucous	mucus	none	lubrication	
Large Intestine	mucous	mucus	none	lubrication	churning peristalsis
Rectum	mucous	mucus	none	lubrication	peristalsis
Anus	mucous	mucus	none	lubrication	defecation

LABORATORY EXERCISE PROCEDURE

Follow the listed procedure to complete the laboratory exercise. Answer the appropriate questions in the laboratory exercise as you proceed. When complete, submit the report sheets to the instructor for grading.

LABORATORY EXERCISES

I. Read the instructions on the report sheet carefully for this section.

II. Carbohydrate Digestion

    A. Specific Objectives

        1. Learn to test for starch in food using iodine.
        2. Learn to test for sugar in food using Benedict's solution.
        3. Demonstrate that saliva digests starch in food to sugar.

    B. Test for Starch

        1. Materials: see first page of this exercise

        2. Procedure
           a. Add 5 drops of iodine to 10 drops of water in a glass (this is your solution).
           b. Add 1/2 tsp cornstarch to 1/4 cup of water and stir.
           c. Add the iodine solution to the cornstarch solution. (A purplish, blue–black color indicates presence of starch.)
           d. Drop iodine onto pieces of bread, potato, and apple. Test rice and oatmeal.
           e. Test 5 other foodstuffs for the presence of starch.
           f. Record all results in the chart on the lab report.
           g. Answer all questions in the lab report.

    C. Test for Sugar

        1. Materials: see first page of this exercise

        2. Procedure
           a. Add a few drops of Benedict's solution to a test tube 1/3 full of water and heat toboiling. The test tube should be placed in the jar with water in the jar. The jar should be set in the saucepan and then heated to boiling. **CAUTION**: The test tube will be hot to the touch and can cause burns. The boiling liquid can escape the tube and cause burns. Be careful!
           b. Add 1/2 tsp. sugar to another tube 1/3 full of water and stir.
           c. Add a few drops of Benedict's solution to the test tubes containing the sugar-water and water only. Heat both test tubes to boiling by placing the test tubes in the jar of water and put the jar in the sauce pan. **Be careful!** (A yellow, gold, green, orange or brick red color after heating indicates the presence of sugar.)

    D. Conversion of Starch to Sugar

        1. Materials: see first page of this exercise

2. Procedure
   a. Chew about 1/4 of a non–sugar cracker thoroughly, so that it is mixed with saliva. Expectorate the cracker–saliva into a test tube. Crumble some of the cracker into another test tube and add water to moisten the cracker. This is the control.
   b. Warm both test tubes by placing the test tubes in a heat–resistant jar. Set the jar in water in the small sauce pan. Carefully warm the sauce pan on the stove but do not allow the water to boil. The water should be kept at about 90° F for 5 minutes.
   c. After 5 minutes, add several drops of Benedict's solution to both test tubes and heat to boiling. **Be careful** as boiling occurs.
   d. Answer all questions in the lab report.

III. Test for Fat

   A. Material: see page one of this laboratory exercise

   B. Procedure

      1. Rub the bacon on a piece of unglazed paper.
      2. Hold the paper up to the light. (If light comes through a translucent spot on the paper, fat is present.)
      3. Do a similar test with a liquid fat such as vegetable oil.
      4. Answer all questions in the lab report.

IV. Complete the discussion questions.

THE DIGESTIVE SYSTEM

Report Sheet 1

Name _____

Student ID # _____

Campus _____

Date _____

I. Using the structure list below, label the drawing of the digestive system. Place the appropriate letter beside the number of the structure or organ. Then match the function (using the *letters*) with the proper structure. Credit will only be given for letter answers. Refer to pages 116 & 120 in the *Photo Atlas* for help. All labels may not be used.

Structure List

a. appendix
b. ascending colon
c. cecum
d. cystic duct
e. descending colon
f. duodenum
g. esophagus
h. gall bladder
i. common hepatic duct
j. ileum
k. jejunum
l. liver
m. pancreas
n. pancreatic duct
o. parotid gland
p. pharynx
q. rectum
r. sigmoid colon
s. stomach
t. sublingual gland
u. submandibular gland
v. transverse colon
w. tongue

Function List

aa. stores bile
bb. produces bile
cc. produces saliva
dd. passageway for food
ee. carries bile to duodenum
ff. positions food for chewing, swallowing
gg. carries bile to gall bladder
hh. digestion of chyme
ii. protein digestion
jj. carries pancreatic juice to small intestine
kk. produce digestive enzymes, hormones
ll. digestion of chyme, absorption
mm. solidification of wastes
nn. water absorption
oo. semi–solid waste storage
pp. no function known
qq. absorption of digestive end–product

(ex.) 1. __o__ __cc__   7. ____ ____   13. ____ ____   19. ____ ____

2. ____ ____   8. ____ ____   14. ____ ____   20. ____ ____

3. ____ ____   9. ____ ____   15. ____ ____   21. ____ ____

4. ____ ____   10. ____ ____   16. ____ ____   22. ____ ____

5. ____ ____   11. ____ ____   17. ____ ____   23. ____ ____

6. ____ ____   12. ____ ____   18. ____ ____

ORGANS OF THE DIGESTIVE SYSTEM

small intestines

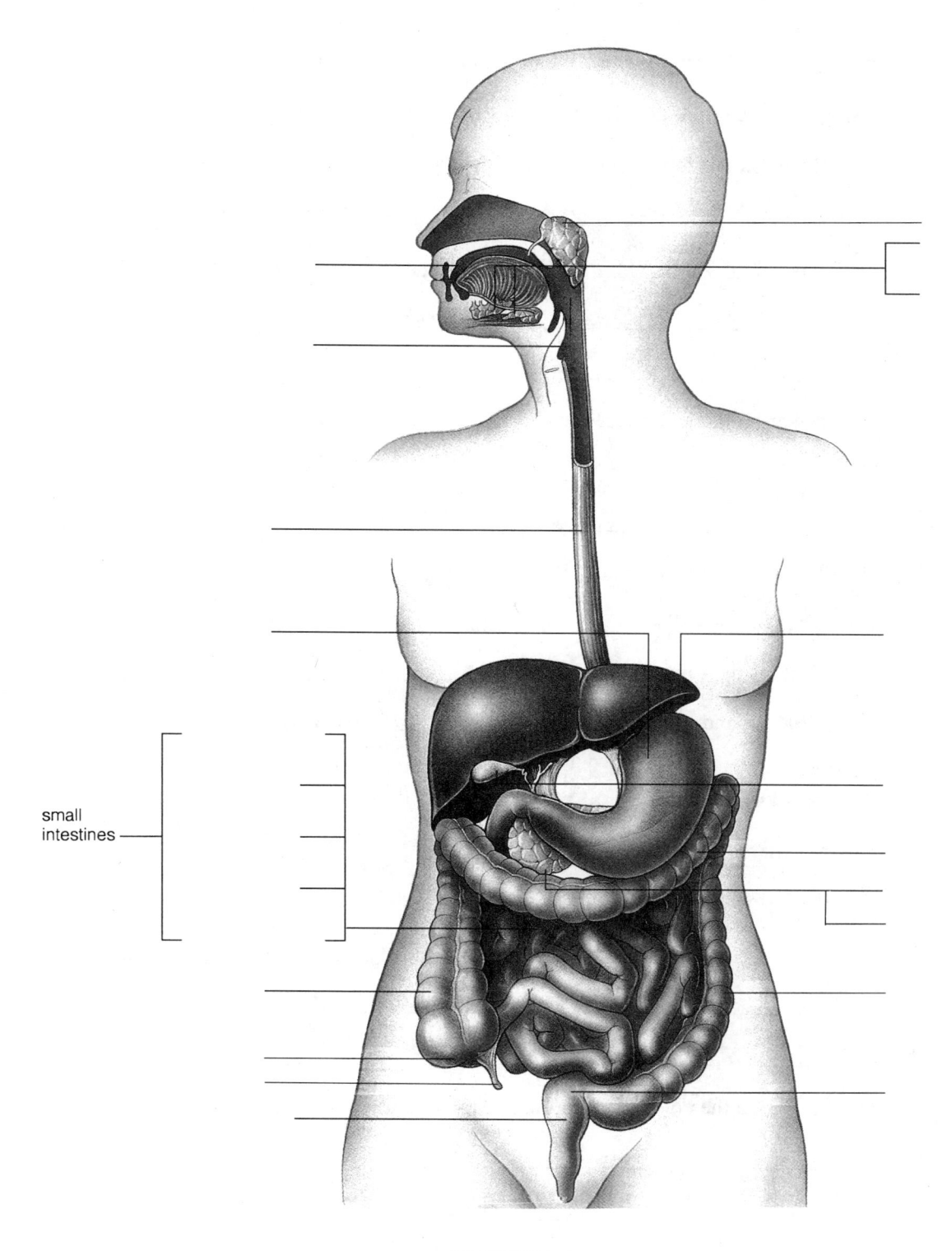

FIGURE 1

The Digestive System Student ID # _____
Report Sheet 2

II. Digestive Organs—Refer to your textbook and the discussion for help

   A. Using your textbook and discussion for help, answer the following questions:

      1. Describe the four layers of the esophagus

         Layer 1 _____

         Layer 2 _____

         Layer 3 _____

         Layer 4 _____

      2. Describe the differences in the mucosa of the esophagus and the stomach

         _____

         _____

      3. What is the function of the chief cells? _____

         The parietal cells? _____

      4. What is the function of a villus? _____

      5. Glucose and amino acids enter _____ Fatty acids enter _____

III. Carbohydrate Digestion

   A. Test for Starch

      1. Describe the test for starch_____

         _____

         _____

         _____

      2. Describe the color of the cornstarch solution before adding iodine._____

         _____

         _____

The Digestive System
Report Sheet 3

Student ID # _____

3. Describe the color of the cornstarch solution after adding iodine. _____

_____

_____

4. Did a color change occur? _____ If so, what did the color change indicate?

_____

_____

5. Fill in the following chart.

Food (last 5 are foods of choice)	Color before Iodine added	Color after Iodine added	Conclusion
bread			
potato			
apple			
rice			
oatmeal			

B. Test for Sugar

1. Describe the test for sugar. _____

_____

_____

_____

The Digestive System Student ID # _____
Report Sheet 4

2. Describe the color of water before and after adding Benedict's solution.

   Before _____

   After _____

3. Describe the color of the sugar–water solution before adding Benedict's solution.

   _____

   After adding Benedict's solution _____

   After heating _____

4. Did a color change occur after heating the sugar–water Benedict's solution mixture?

   _____ If so, what does this indicate? _____

   _____

5. What precaution must be observed during this experiment? _____

   _____

   _____

C. Conversion of Starch to Sugar

   1. Why is saliva added to the soda cracker?

      _____

      _____

   2. Is sugar indicated in the tube with saliva after heating? _____

      _____

      _____

   3. Is sugar indicated in the control tube? _____ Why is there no sugar found in the control?

      _____

      _____

246 Exercise 23

The Digestive System
Report Sheet 5

Student ID # _____

A. Describe the test for fat. _____

_____

_____

B. Describe the appearance of the paper after using the bacon. _____

_____

vegetable oil _____

IV. Discussion Questions

1. List and describe the basic steps in digestion.

   a. _____

   b. _____

   c. _____

   d. _____

   e. _____

2. List and define the physical or mechanical processes of digestion.

   a. _____

   b. _____

   c. _____

3. Define chemical digestion _____

_____

_____

The Digestive System　　　　　　　Student ID # _____
Report Sheet 6

4. List, in order, the major parts of the digestive tract.

   a. _____

   b. _____

   c. _____

   d. _____

   e. _____

   f. _____

   g. _____

   h. _____

5. List and describe the function of the accessory digestive organs.

   a. _____

   b. _____

   c. _____

   d. _____

   e. _____

6. Identify the following:

   a. bolus _____

   b. chyme _____

7. List the breakdown products of

   a. carbohydrates _____

   b. lipids _____

   c. proteins _____

248　Exercise 23

24 NUTRITION

LESSON OBJECTIVES

Upon completion of this laboratory exercise the student will be able to:

1. List the nutrients necessary for a balanced diet.

2. List foods which contain the necessary nutrients for a balanced diet.

3. Discuss the four basic food groups.

4. Perform a diet analysis.

5. Determine energy expenditures in calories for a day.

6. List the functions, sources and deficiency disorders for carbohydrates; fats; proteins, water; Vitamins A, D, K, B, C; and the minerals calcium, iron, and iodine.

7. Compete the lab report sheet.

MATERIALS NEEDED

1. Laboratory Manual
2. Textbook

PREPARATION

It is essential to the successful completion of this laboratory exercise that the entire exercise including the report sheet be read before attempting to complete this laboratory. Also read the appropriate chapter(s) in your textbook.

DISCUSSION

Food is what is eaten and **nutrition** is how the body uses food. Individual components in foods that are required by the body are termed **nutrients**. Although a variety of factors may influence the amount and type of foods eaten, from a cellular viewpoint, the basic purposes of food are to supply structural materials and to provide energy.

A variety of nutrients are required to meet all of the needs of the human body. The nutrients necessary for the body's needs can be classified according to the function they serve. The Nutrient Chart (Table I) provided with the laboratory exercise lists the nutrient types, functions, some common sources, and some deficiency disorders.

Foods can be divided into four major food groups based on their similarity in nutrient content. These four food groups include the **meat** group, the **vegetable–fruit** group, the **milk** group and the **bread–cereal** group. The four food groups and the foods included in each are shown in Table II.

The energy obtained from nutrients may be used by the body's cells in many different ways. The energy can be converted to heat to help maintain the body temperature. A large percentage of the energy is converted into adenosine triphosphate (ATP). ATP is a chemical used by the body to power all of its functions. It provides energy for breathing, digestion, circulation, walking, working, talking, relaxing and making the structural components of the body.

The human body is in energy balance when the food ingested provides energy equal to the energy needs of the individual. A **calorie** is a unit of heat energy and is used to refer to the amount of energy derived by the individual from the metabolism of foods. One gram of carbohydrate or protein provides 4 calories and one gram of fat provides 9.3 calories (1 pound = 454 grams). The caloric content of most foods has been determined and is available in numerous publications.

In relation to weight gain or loss, it is interesting to note that one pound of body weight is equal to 3500 calories. Therefore, if more calories are ingested than the individual uses, weight would be gained.

Weight can be lost by increased activity which uses up calories or by the taking in of fewer calories. It is generally recommended that the daily caloric intake should not be less than 1200 calories to help insure that all the necessary nutrients needed by the body are ingested. Further, the food should be selected with extreme care.

The amount of energy required to perform certain activities has been estimated and is summarized in Table III. Using the tables and the work sheet provided it is possible to calculate a personalized estimate of the energy expenditure during a 24–hour period.

An analysis of the diet for a 24–hour period is also possible and the necessary information to complete this analysis is contained in this lab. A diet analysis will help identify the nature, quantity and sources of food eaten to determine whether the food consumed meets the nutritional needs of the body.

TABLE 1

Nutrient	Sources	Function	Deficiency Disorders
Carbohydrates	Complex: cereals, grains, rice, fruits, vegetables Simple: anything containing sugar (sucrose syrup or honey)	Energy	Incomplete fat oxi–dation; formation of acid–reacting ketone bodies in abnormal amount; excessive breakdown of body protein; dehydration
Fat	Fats and oils found in butter, salad oil, margarine, nuts, meat, whole milk, cream, cheese	Energy, Flavor and Variety. Also provide necessary fatty acids	Reduced absorption of Vitamin E from the digestive tract, dermatitis and growth impairment in children
Protein	Meat, poultry, fish, milk, eggs, peas and beans	Body fluids, muscle, skin, all body tissues	Impaired growth
Water	Beverages, fruits and vegetables	Solvent for body nutrients and wastes	Loss of fluid equilibrium
Vitamin A	Whole milk, green & yellow vegetables	Skin and eyes	Night blindness; dry brittle epithelia of skin, respiratory system and urogenital tract
Vitamin D	Sunshine, fortified milk	use of calcium, strong bones	Rickets in children; osteomalacia in adults
Vitamin D	Vegetables	Blood	Slow blood clotting and hemorrhage, hemophilia
Vitamin E	Vegetable oils	DNA synthesis	None known in man
Vitamin B (a group of about 6)	Meats & whole grains	Nervous system	Beriberi; dermatitis; impairment of antibody synthesis; anemia; pellagra
Vitamin C	Citrus fruits	Immunity, connective substance, healing	Scurvy; arthritic pain; loose teeth; anemia; slow healing of wounds
Minerals: calcium, magnesium phosphorus	Milk, dark green vegetables	Blood clotting, teeth, muscle contraction	Weak bones none none
Iron	Meats	Hemoglobin in red blood	Reduced oxygen–carrying ability of red blood cells
Iodine	Seafood, iodized salt	Thyroid function	Goiter
Copper, Zinc, Selenium, Chromium		Trace elements needed in minute quantities	Only needed in small amounts easily obtainable from good diet practices

TABLE II

FOUR BASIC FOOD GROUPS

Food Group	Foods Included	Recommended Servings	Serving Size
Vegetable–Fruit	All vegetables and fruits (contains sources of Vitamins C and A)	A dark green or deep yellow vegetable at least every other day. Citrus fruit daily. 4 servings daily	1/2 cup cooked vegetables; fruits—1 cup or 1 medium-sized
Cereal–Bread	Whole–grain, enriched or restored (check labels)	4 or more servings daily	1/3—1 cup cereal; bread—1 slice
Meat	Beef, veal, lamb, pork, poultry, fish, shellfish, eggs, dry beans, peas, nuts, peanuts, peanut butter	2 or more servings daily	3 oz. cooked
Milk	Milk, cheese, ice cream, yogurt	Children—3 or more cups Teenagers—4 or more cups Adults 2 or more cups Pregnant women—3 or more cups Nursing mothers—4 or more cups	1 cup = 1 serving

TABLE III

ENERGY EXPENDITURES CHART*

Activity	Rate (calories per kilogram per hour)
Sleeping	1.0
Sitting	1.1
Personal care (dressing, etc.)	2.2
Walking (moderately)	3.2
Climbing stairs	9.0
Eating	1.3
Watching TV, listening to radio, talking	1.3
Studying, sitting in class	1.3
Running	12.0
Swimming	7.0
Shopping	4.0
Cooking, dusting, setting table, washing dishes	2.5
Sweeping, ironing	1.9
Bedmaking, vacuuming and moving furniture	3.8
Scrubbing floors, shopping with a heavy load	4.0
Office work, selling, lab technician, hairdressing, hospital work, repairing autos, truck driving	3.0
Carpentry	5.0
Farming, partly mechanized	5.0
General labor	5.0
Garbage collection, postal delivery	7.5
Playing cards, musical instruments	2.0
Archery, bowling, golf, sailing, volleyball	3.0
Badminton, canoeing, cycling, dancing, gardening, gymnastics, horse riding, skiing, tennis	5.0
Football, boxing, basketball, track and field	6.0

*These figures are approximations of energy expended in the various activities listed. The vigor with which any individual enters into an activity would alter the results.

The above values were interpreted and compiled from R. Passmore and J.V.G.A. Durnin, "Human Energy Expenditure," *Physiological Reviews*, 35 (October 1955), 801 and J.V.G.A. Durnin and R. Passmore, *Energy, Work and Leisure*, London: Heinmann Educational Books, LTD., 1967.

LABORATORY EXERCISE PROCEDURE

Carefully read the directions for each exercise. Complete the exercises, recording the required information on the report sheets. Note paper may be used to gather the information, assemble it and then record it on the report sheet. When complete, submit the report sheets to the instructor for grading.

LABORATORY EXERCISES

I. An analysis of food consumption for one day is to be completed in this exercise including a food consumption tabulation, four basic food group chart, and a diet rating scale.

   A. Food Consumption Tabulation—Keep a record of *all* the food consumed in one day (a 24–hours period). Record this information on the Report Sheet Section IA. Keep a record of the serving size of each food such as steak—4 oz., of rice—1 cup. Determine the caloric content of each food. Most canned, frozen, bottled or packaged foodstuffs have the caloric content on the label. If foods are consumed that don't have the caloric content given, refer to the Nutrition Section of an encyclopedia, a calorie counter book (numerous inexpensive editions are available) or examine copies of nutrition books in the library.

   B. After completing Section IA—the Food Consumption Tabulation—continue the analysis of the day's diet by determining the servings per food group using the chart provided on the Report Sheet Section IB and Table II in this laboratory exercise.

   C. A diet rating scale will complete the diet analysis section of this lab. Refer to Section IB to aid in completion of the diet rating scale. (Caution: Do not forget to calculate the Final Score on this section.)

II. It is possible to calculate the energy expenditure for a day using Table III and the chart on the report sheet (IIA). List each activity participated in for a 24–hour period. Then determine the amount of time spent in each activity. By multiplying the time spent by the rate (calories per kilogram per hour) the calories per kilogram spent on a particular activity will result.

   EXAMPLE:
   If 1 1/2 hours is spent eating in a 24–hour period, refer to Table III and see that eating required 1.3 calories per kilogram per hour.
   1-1/2 hours × 1.3 calories/kilogram/hour = 1.95 calories/kilogram

   Another example is given on the Report Sheet.

   B. The total daily energy expenditure can be determined using the totals from Section IIA. Follow the report sheet carefully. You need to know your body weight before starting this section.

III. Using the information collected in Sections I and II, answer the questions on the report sheet under Section IIIA. Fill in the chart in Section IIIB using Table I, the Nutrient Chart and Section IB of the Report Sheet.

IV. Vitamin Deficiency Disorders

   Using your textbook for help, answer the questions on the Report Sheet, Section IV.

NUTRITION

Report Sheet 1

Name _____

Student ID # _____

Campus _____

Date _____

I. Diet Analysis

Objective: To identify the nature, quantity and source of food eaten to determine whether the ingested food meets nutritional requirements.

A. Food Consumption Tabulation

Date	Food	Food (Serving Size)	Calories*
Breakfast			
Mid–morning			
Lunch			
Afternoon			
Dinner			
After Dinner			

Total Calories _____
(also record on Section IIB part 3a)

*Most canned, frozen, bottled or packaged foodstuffs have the calorie content on the label. If you consume food that do not have the calorie content given, refer to the Nutrition Section of an encyclopedia, a calorie counter book or examine copies of nutrition books in the library.

Nutrition Student ID # _____
Report Sheet 2

B. Four Basic Food Groups

After completing Section IA—the Food Consumption Tabulation—continue the analysis of the diet by determining the servings per food group using the chart below.

Food Group	Specific Food Eaten	Serving Size
Vegetable—Fruit		
Cereal—Bread		
Meat		
Milk		
Sweets*		
High Fat Foods (coconut oil, chocolate)		
High Sodium Nitrate Foods (bacon, ham, sausage)		
Miscellaneous		

* More than 2 sweets or empty calorie snacks per day is unsatisfactory

Nutrition
Report Sheet 3

Student ID # _____

C. Diet Rating Scale—Refer to IB to aid in the completion of this section.

Number of food groups fulfilled	_____	× 25 =	_____
Number of vegetables (see totals from Food Consumption Tabulation Sheet	_____	× 5 =	_____
Number of fresh unprocessed vegetables	_____	× 2 =	_____
Number of fruit	_____	× 5 =	_____
Number of fresh unprocessed fruit	_____	× 2 =	_____
Number of protein foods	_____	× 5 =	_____
Number of non–meat protein foods	_____	× 2 =	_____
Number of grains and breads	_____	× 5 =	_____
Number of whole grain foods	_____	× 2 =	_____

TOTAL #1 _____

Number of sweets	_____	× 5 =	_____
Number of high fat foods	_____	× 5 =	_____
Number of foods containing sodium nitrate	_____	× 5 =	_____

TOTAL #2 _____

FINAL SCORE: Total #1 _____

Total #2 _____

_____ FINAL SCORE *

* If Total #1 is 10 points higher than Total #2 the diet is Good.

If Total #1 is 5 points higher than Total #2 the diet is Satisfactory.

If Total #1 equals Total #2, the need to examine eating habits exists—too many sweets, high fat foods and/or foods high in sodium nitrate are being consumed.

If Total #2 is higher than Total #1, the diet for the day was not compatible with good nutritional practice.

Nutrition
Report Sheet 4

Student ID # _____

II. Energy Expenditure For A Day

Objective: To determine the calories expended during daily activities and relate this nformation to calories ingested.

A. Energy Expenditures per Activity Record

List each activity participated in for a 24–hour period and determine the time spent in each activity. Then multiply the Total Time Spent by Calories/Kg/Hour to obtain Calories /Kg.

EXAMPLE:

Activity	Total Time Spent (hours)	Calories/Kg/Hour (See Table III)	Calories/Kg for activity time
Sitting	15 min = 1/4 hr	1.1	275

Activity	Total Time Spent (hours)	Calories/Kg/Hour (See Table III)	Calories/Kg for activity time

TOTALS Time (24 hours or 1440 minutes)

(Also record this figure on Section IIB, number 2)

Nutrition Student ID # _____
Report Sheet 5

B. Total Daily Expenditure from Activity Record

1. Divide the body weight in pounds (lbs.) by 2.2 to obtain the weight in kilograms.

   $\dfrac{\underline{\qquad} \text{ lbs.}}{2.2 \text{ Kg/lbs}}$ = _____ Kg.

2. Multiply the body weight in Kg. by total calories per Kg. (from IIA).

   _____ Kg × _____ Cal./Kg = _____ Calories (Enter on the line
   Weight   provided in 3b)

3. From IA determine the total calories ingested and fill in the information below and subtract b from a.

   a. Calories ingested   _____

   b. Calories expended _____

   _____ *

* If the answer is a positive number, more calories were ingested than used. If the answer is negative, more energy in calories was expended than ingested and weight would be lost over a period of time if this pattern continued.

III. Using the information collected under Sections I and II answer the following questions.

A. Did the diet for a day contain foods from the four basic food groups? (See IIB)

_____ Were the serving sizes adequate to provide the minimum recommended

amounts of these foods? _____ If not, what was missing (either basic food

group(s) or serving sizes)? _____

B. Refer to Table I, the Nutrient Chart, and IB for the following analysis.

Nutrient	Function	Foods eaten which furnish this nutrient
Carbohydrate		
Fats		

Nutrition   259

Nutrition Student ID # _____
Report Sheet 6

Nutrient	Function	Foods eaten which furnish this nutrient
Proteins		
Water		
Vitamins:		
A		
D		
K		
B		
C		
Minerals:		
Calcium		
Magnesium		

Nutrition Student ID # _____
Report Sheet 7

Nutrient	Function	Foods eaten which furnish this nutrient
Phosphorus		
Iron		
Iodine		
Copper, Zinc, Selenium, Chromium:		

IV. Nutritional Disorders

    A. Define the following disorders including the cause(s):

        1. Xerophthalmia _____

           _____

           Cause_____

           _____

        2. Pellagra _____

           _____

           Cause_____

           _____

Nutrition
Report Sheet 8

Student ID # _____

3. Rickets _____
   _____

   Osteomalacia _____
   _____

   Cause_____

4. Goiter _____
   _____

   Cause_____
   _____

B. Answer the following questions

1. What effect does Vitamin B_2 deficiency have on the skin? _____
   _____
   _____

2. What effect does Vitamin D deficiency have on bone? _____
   _____
   _____

3. What effect does potassium deficiency have on heart muscle? _____
   _____
   _____

4. What effect does iron deficiency have on blood? _____
   _____
   _____

5. What effect does Vitamin B_1 deficiency have on heart muscle? _____
   _____
   _____

25 THE EXCRETORY SYSTEM

LESSON OBJECTIVES

Upon completion of this laboratory exercise the student will be able to:

1. Identify and describe the function of the major parts of the excretory system.
2. Identify and describe the function of the major parts of the urinary system.
3. Identify the major parts of the kidney.
4. Identify and describe the function of the major parts of the nephron.
5. Perform the experiments in the exercise.
6. Answer all questions in the lab report.

MATERIALS NEEDED

1. Laboratory Manual
2. Textbook
3. Tap water
4. Plastic bag (such as a sandwich bag)
5. Salt water (add table salt to water)
6. Cotton balls
7. Urine sample
8. Test tubes*
9. Cobalt chloride paper*
10. Silver nitrate*
11. Distilled water*
12. *Photo Atlas for Biology*

   * These items can be found in the lab kit.

PREPARATION

Read the discussion which follows carefully before attempting to complete the exercise. Also read the appropriate chapter(s) in your textbook.

DISCUSSION

The primary excretory organs of the body are the kidneys, skin, lungs, and intestines. The **kidneys** are a part of the urinary system and function to excrete or remove from the body nitrogen waste products such as urea in urine, toxins, and water. As a secondary function, the kidneys help regulate the balance of acids and bases in the body. Special glands of the skin, the **sweat glands**, serve as excretory organs. They remove water, mineral salts, and when necessary, small amounts of nitrogen wastes. The sweat glands also function to help cool the body due to the evaporation of the water excreted by the glands onto the skin surface. As excretory organs, the **lungs** excrete carbon dioxide, CO_2, produced as a waste product during cellular respiration, and water. In addition the lungs serve as the means by which oxygen is taken into the body and as the means, due to the excretion of CO_2, by which acid–base balance is maintained. The **intestines** excrete the undigested food products and some metabolic waste materials such as bile pigments.

The major organs of the excretory system are the urinary organs. They include the kidneys, ureters, bladder, and urethra. The **kidneys** are bean-shaped organs lying against the back abdominal wall and held in place by fat cells and fascia, a membrane covering. The inner part of the kidney is called the **medulla** and the outer part is called the **cortex**. The functional unit of the kidney is the **nephron**, which consists of a glomerulus, its tubules, and its blood supply.

The **kidney** is a bean-shaped organ, indented on one side. The indented area is the area where the renal arteries enter the kidney carrying oxygenated blood to the kidneys and the renal vein leaves the kidneys removing blood from the kidney after its cleansing. The ureter also leaves the kidney at this point. The functional unit of the kidney is the **nephron**, which is located in the cortex and outer medulla region. The nephron consists of a tuft of capillaries called the **glomerulus** surrounded by a cup-shaped structure called **Bowman's capsule** or the **glomerular capsule**. Blood is filtered from the glomerulus into Bowman's capsule. Blood filtrate includes water, glucose (sugar), salts, nitrogen wastes, and other molecules in lesser amounts. Attached to Bowman's capsule is the tubular system of the nephron. The first part of the tubular system is the **proximal convoluted tubule** where the majority of the water and the glucose filtered from the glomerulus is reabsorbed into the blood supply. Attached to the proximal convoluted tubule is the **loop of Henle** which is responsible for the concentration or dilution of urine. The **distal convoluted tubule**, attached to the loop of Henle, functions as the site of selective water and ion (salt) return to the blood supply. Return of water and salt to the blood is on the basis of need and is controlled by hormones. The hormone **ADH** (antidiuretic hormone—against water loss) controls water absorption from the distal tubule and the hormone **aldosterone** controls salt ion absorption. ADH is produced in the hypothalamus and aldosterone in the adrenal gland. The distal convoluted tubules of many nephrons are attached to the **collecting duct** or tubule. From this point on the blood filtrate, now called urine, remains unchanged as it passes out of the kidney on its way out of the body. The collecting duct empties into the **calyx** which empties into a large cavity called the **renal pelvis**. The renal pelvis empties into the **ureter**. The ureter extends from the kidney and drains it into the urinary bladder. The **bladder** serves as a reservoir for urine. When it fills, it expels urine from the body by way of the **urethra**. In the female, the urethra lies behind the pubic bone, in front of the vagina, and serves to expel urine from the body. In the male, the urethra extends through the prostate gland and the penis. It serves to expel urine from the body as an excretory function and expel semen from the body as a reproductive function. Expulsion of urine and semen occur at different times. They cannot occur at the same time due to the closing off of the reproductive pathway during voiding (urine expulsion) and the closing off of the urinary tract during ejaculation (semen expulsion).

Urine, the kidney filtrate, consists of approximately 95% water in which are dissolved nitrogen wastes from protein breakdown, mineral salts, toxins, and other miscellaneous components. Urine normally has a straw-yellow color which may be darker when the urine is concentrated or lighter to the point of being clear when the urine is diluted. Abnormal constituents of urine include glucose (sugar), ketones, protein, blood, bilirubin, and urobilinogen. Normally no detectable glucose is present in urine, but small amounts may appear after eating a heavy meal or in conjunction with emotional stress. The chief pathological state leading to glucose in the urine is diabetes mellitus, although renal or liver damage may also be a cause. During fasting, starvation, diabetes mellitus, and other abnormalities of carbohydrate metabolism, ketone bodies may be present. Normally ketones are not present. Normal urine contains traces of protein, but it does not give a positive reaction with the usual clinical tests for protein. In most cases, when protein is found in the urine in large quantities, renal disease such as acute glomerulonephritis is the cause. Protein in the urine may also result from cardiac abnormality, anemia, liver disease, fever, or certain cancers. Anything more than a trace of protein in the urine is considered clinically significant. Blood is detected in the urine in the form of hemoglobin. Severe infections, burns, some poisonings, renal disease and even female menstruation may all result in blood being present in the urine. Normally no bilirubin is detectable in urine; however, even trace amounts of bilirubin in urine are sufficiently abnormal to require further investigation since its presence indicates a liver problem. Urobilinogen is normally excreted in the feces, but some liver problems, malaria, and infectious mononucleosis may cause its appearance in urine.

Several commercial urine test strips are available. They serve to provide a rough estimate of the amount of these abnormal constituents that might be present. These test strips can be purchased from most drug stores and include the "Multistix," "Hema–Comistix," and "Ketostix'""as well as other names.

LABORATORY EXERCISE PROCEDURE

Carefully complete each exercise as indicated and record the required information on the report sheets. When complete, submit the report sheets to the instructor for grading.

LABORATORY EXERCISES

I. Study the drawings of the excretory system. Refer to the discussion section and the text book to match each structure with its function. (Refer to pp. 119 and 121 in the *Photo Atlas* for help.)

II. A. Objective: Determine if skin and lungs aid the kidneys in the removal of similar and different waste chemicals from the body.

B. Materials: see page one of the exercise

C. Explanation of chemical reactions:

1. The presence of water may be detected by using cobalt chloride paper. A blue strip of cobalt chloride will turn pink in the presence of moisture (water).

2. The presence of salt is confirmed when a milky white condition is formed by the interaction of salt solution and silver nitrate.

D. Procedure

1. Test for water loss through the lungs and skin

a. In the chart on the lab report sheet (II) record the color of dry cobalt chloride paper. **NOTE**: Cobalt chloride paper should be blue. If the paper is pink, it hasbeen exposed to moisture. Before proceeding, heat the paper

in a microwave oven or conventional oven set at 350° F until the paper regains its blue color.
    b. Place a drop of water on a small piece of paper. Record the results on the report sheet.
    c. Place a piece of the dry cobalt chloride paper in a small plastic bag (a sandwich–size bag is ideal). Exhale into the bag, tightly close the top of the bag, and observe the paper for a color change. Record any change in the color of the paper on the report sheet. **CAUTION**: Plastic bags can be dangerous—use only as directed.
    d. Lay a piece of the cobalt chloride paper in the palm of the hand. Record any color change which occurs.

III. Test for salt loss

The Scientific method lab discussed the need for controls in experiments. The first step in testing for salt loss will be to set up controls.

    A. Controls: Add distilled water to a test tube until it is 1/4 full. Add salt water to a second test tube until it is 1/4 full. To each tube, add one drop of silver nitrate solution. (**CAUTION**: Handle the silver nitrate carefully as it will stain your skin, clothes, etc.) Record what happens in each tube on the report sheet.

    B. Next place a small ball of cotton (about the size of a pea) in a test tube 1/4 full of distilled water.

    C. In a second test tube 1/4 full of distilled water, place a piece of cotton which has been rubbed across the palm of the hands 10–12 times.

    D. Add one drop of silver nitrate to each of the tubes. Record the observations on the eport sheet.

    E. Fill a test tube 1/4 full of fresh urine. Add one drop of silver nitrate to it. Record he observations on the report sheet.

IV. Answer the questions on the report sheet. Refer to the discussion and to the experiment.

FIGURE 1

Organs of the human urinary system and their functions. The two kidneys, two ureters, and urinary bladder are located *outside* the membranous lining (peritoneum) of the abdominal cavity.

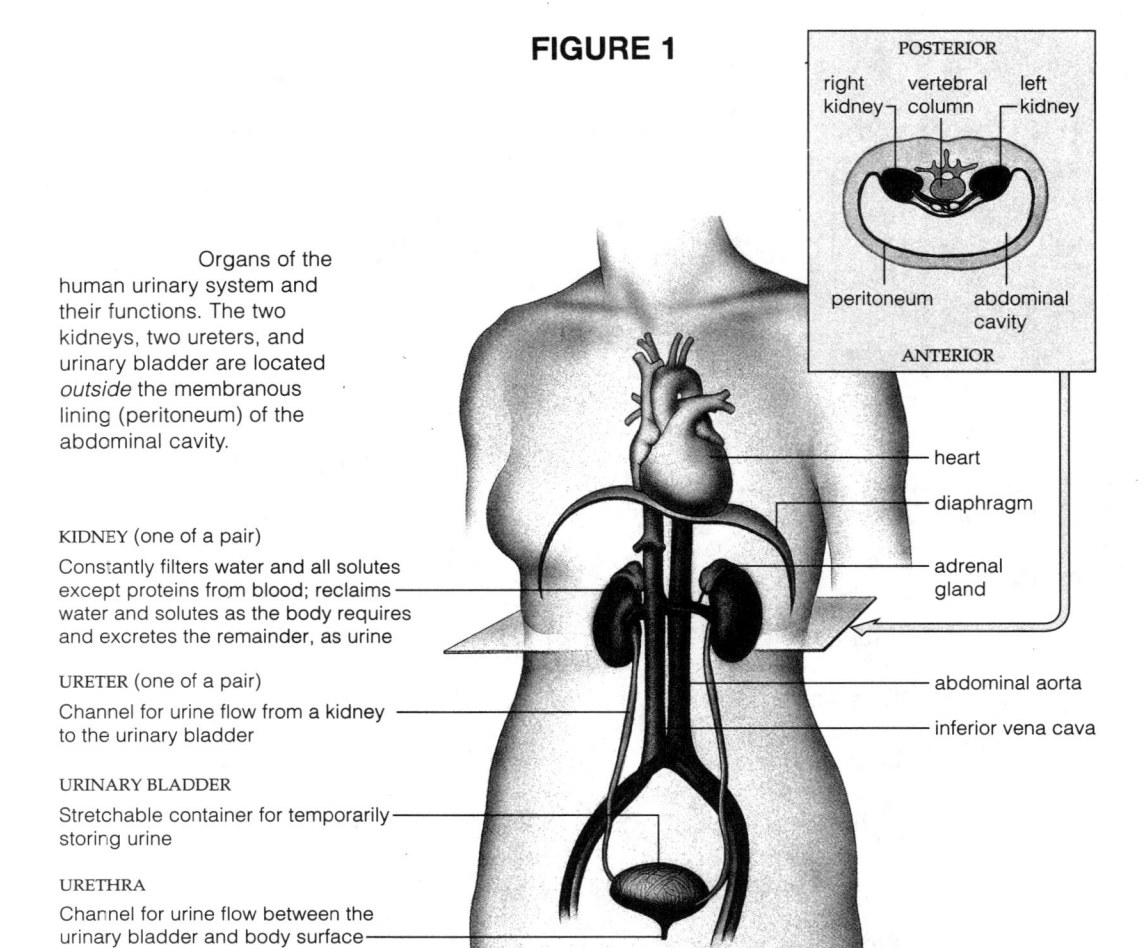

KIDNEY (one of a pair)
Constantly filters water and all solutes except proteins from blood; reclaims water and solutes as the body requires and excretes the remainder, as urine

URETER (one of a pair)
Channel for urine flow from a kidney to the urinary bladder

URINARY BLADDER
Stretchable container for temporarily storing urine

URETHRA
Channel for urine flow between the urinary bladder and body surface

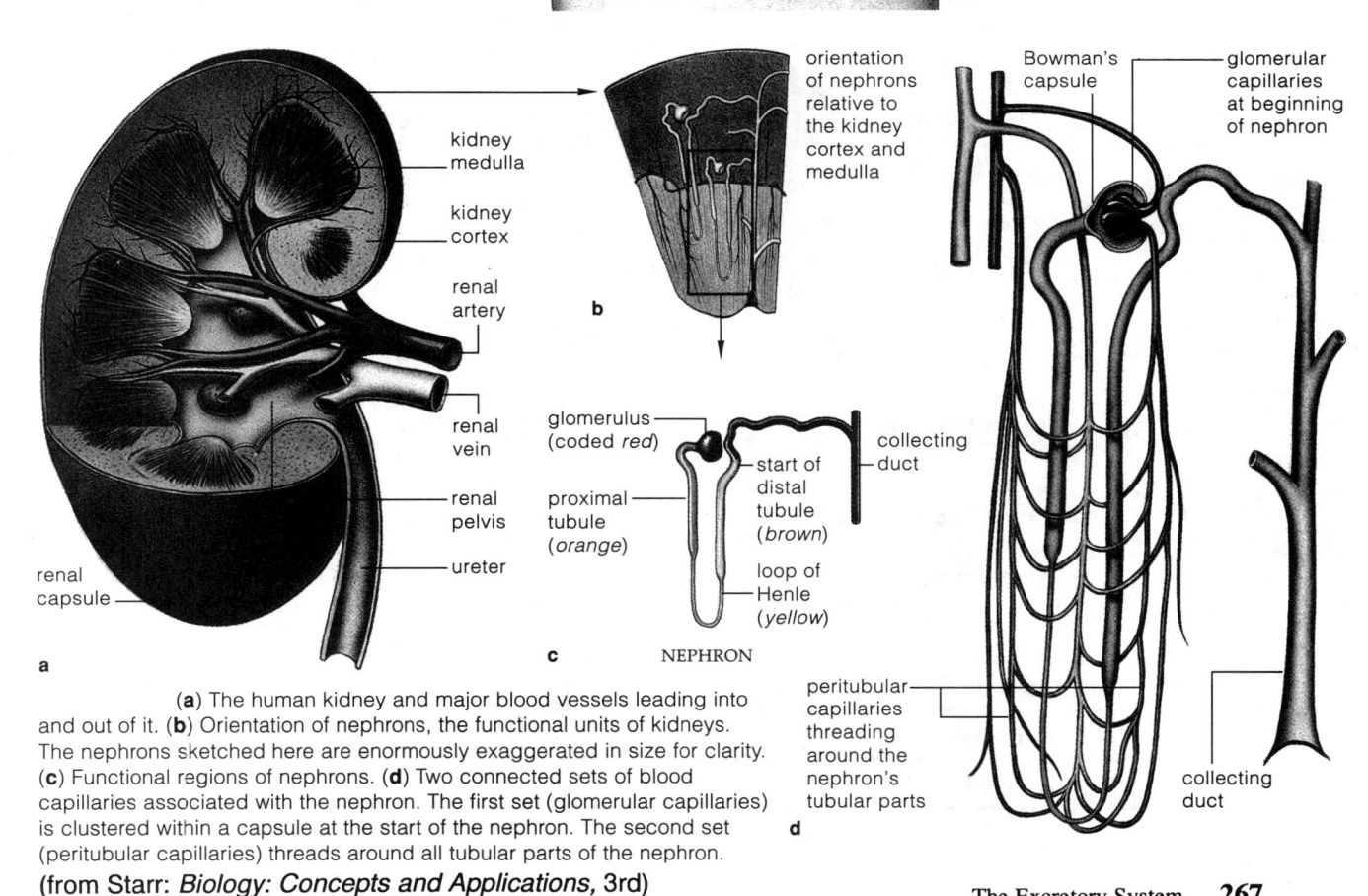

POSTERIOR
right kidney — vertebral column — left kidney
peritoneum — abdominal cavity
ANTERIOR

heart
diaphragm
adrenal gland
abdominal aorta
inferior vena cava

kidney medulla
kidney cortex
renal artery
renal vein
renal pelvis
ureter
renal capsule

a

orientation of nephrons relative to the kidney cortex and medulla

b

glomerulus (coded *red*)
proximal tubule (*orange*)
start of distal tubule (*brown*)
loop of Henle (*yellow*)
collecting duct

c NEPHRON

Bowman's capsule
glomerular capillaries at beginning of nephron
peritubular capillaries threading around the nephron's tubular parts
collecting duct

d

(**a**) The human kidney and major blood vessels leading into and out of it. (**b**) Orientation of nephrons, the functional units of kidneys. The nephrons sketched here are enormously exaggerated in size for clarity. (**c**) Functional regions of nephrons. (**d**) Two connected sets of blood capillaries associated with the nephron. The first set (glomerular capillaries) is clustered within a capsule at the start of the nephron. The second set (peritubular capillaries) threads around all tubular parts of the nephron.

(from Starr: *Biology: Concepts and Applications,* 3rd)

THE EXCRETORY SYSTEM

Report Sheet 1

Name _____

Student ID # _____

Campus _____

Date _____

I. Match the structures of the excretory system listed below with the correct function. Place the correct letter of the function in the blank preceding the appropriate structure.

Structure

_____ 1. Glomerulus
_____ 2. Bowman's Capsule
_____ 3. Loop of Henle
_____ 4. Collecting Tubule
_____ 5. Distal Convoluted Tubule
_____ 6. Proximal convoluted Tubule
_____ 7. Renal Vein
_____ 8. Renal Artery
_____ 9. Kidney
_____ 10. Ureter
_____ 11. Urinary Bladder
_____ 12. Urethra

Function

a. carry blood from kidney
b. filtration
c. secretes urine from the body
d. eabsorption of water, ions, glucose
e. ilter blood
f. filtration, bring blood to nephron
g. carry blood to kidney
h. reservoir for urine
i. selective reabsorption of water and ions
j. carry urine from kidney to bladder
k. water reabsorption, passageway for for urine
l. concentrate or dilute urine

II. C. Results of Procedures:

1. Test for water loss

Color of cobalt chloride paper

	Before	After
Drop of water		
Air from lungs		
Palm of hand		

268   Exercise 25

The Excretory System
Report Sheet 2

Student ID # _____

III. Test for salt loss

Sample	White Cloudy Appearance Yes	No
Distilled water		
Salt water		
Cotton and distilled water		
Cotton rubbed on palm + distilled water		
Urine		

IV. Questions (Refer to the discussion and to the experiments)

　A. What color change in the cobalt chloride indicated the presence of water?

　_____

　B. Does exhaled air contain water? _____ How do you know? _____

　_____

　C. Does the skin excrete salt? _____ How do you know? _____

　_____

　D. Do the kidneys excrete anything besides salt? _____ If so, what? _____

　_____

　_____

　E. List the primary excretory organs of the body and the substances(s) each secretes.

　　　　Organ　　　　　　　　　　　　　Excreted Substances

　　1. _____　　_____

　　2. _____　　_____

　　3. _____　　_____

　　4. _____　　_____

The Excretory System Student ID # _____
Report Sheet 3

F. Describe the function of the following:

1. ADH _____
   _____

2. Aldosterone_____
   _____

G. List the diseases or conditions which may cause abnormally high amounts of the following substances in the urine. (Refer to the discussion.)

   Blood: _____
   _____

   Bilirubin: _____
   _____

   Ketones: _____
   _____

   Glucose: _____
   _____

   Proteins: _____
   _____

26 THE NERVOUS SYSTEM

LESSON OBJECTIVES

Upon completion of this laboratory exercise the student will be able to:

1. Identify and describe the parts of the neuron using the following terms: axon, dendrite, cell body, myelin sheath, and node of Ranvier.

2. Identify and give the function of the following parts of the brain: cerebrum, cerebellum, thalamus, hypothalamus, midbrain, pons, and medulla.

3. List, in order of their occurrence, the coverings of the brain and spinal cord.

4. Identify the following parts of the spinal cord: gray matter, white matter, motor root, sensory root, and sensory ganglion.

5. Describe what is meant by a reflex arc.

6. Perform the various reflex arc experiments.

MATERIALS NEEDED

1. Laboratory Manual
2. Textbook
3. *Photo Atlas for Biology*
4. Ruler

PREPARATION

Read the discussion that follows carefully and the appropriate chapter(s) in your textbook before completing this exercise.

DISCUSSION

The human nervous system is composed of specialized tissues capable of the complex integrated functions of excitability and conductivity. Due to these functions, humans are aware of their environment and can respond to environmental changes and pressures successfully. For ease of study, the nervous system is divided into two parts: the **central nervous system** (CNS) and the **peripheral nervous system** (PNS).

The brain and spinal cord compose the central nervous system while the peripheral nervous system includes 12 pairs of cranial nerves and 31 pairs of spinal nerves plus the branches from these 43 pairs of nerves. The peripheral nervous system conducts messages from sensory receptors to the central nervous system where the messages are processed to select the appropriate response. The response message is carried from the central nervous system by the peripheral nervous system to effectors (muscles and glands). Receptors include those for touch, smell, taste, temperature, vision, hearing, and balance.

The peripheral nervous system can be further divided into the **somatic nervous system** which goes to the skin, skeletal muscles and joints; and the **autonomic nervous system** which controls the actions of all internal organs and consists of the sympathetic and parasympathetic nerves. Nerves of the sympathetic system send nerve impulses speeding up the action of an organ or gland. Sympathetic impulses prepare the body for a "fight or flight" response to a stimulus. Parasympathetic nerves slow down the action of an organ or gland, producing a "resting and digesting" response.

The mechanism of nerve action involves receiving a stimulus and then responding to it. The stimulus is received by receptors, sensory neuron endings that receive messages (stimuli) from the internal and external environment. The message is then carried by the sensory neuron to either a connecting neuron (called an internuncial neuron) or a motor neuron. Ultimately the message goes to the motor neuron where the motor neuron endings carry the message to either a muscle or gland for action. The nerve impulse is actually a change in electrical charge on the surface or membrane of the neuron. This electrical charge change is the same regardless of the type of stimulus or its destination.

The functional unit of the nervous system is the **neuron**. (See Figure 1.) It consists of nerve fibers and a cell body. The fibers are called **dendrites** and **axons**. A given multipolar neuron would have many dendrites and a single axon. Nerve impulses would travel from the dendrites to the cell body to the axon. Only the axon can transmit the impulse to another neuron or some other effector. Nerve impulse transmission is via a chemical transmitter substance. That substance bridges the gap between the neuron and its target. The gap between two neurons is called a **synapse**. The gap between a neuron and a muscle fiber is called the **myoneural** or **neuromuscular junction**. Once the nerve impulse passes to another neuron or some effector, the original neuron returns to its original state to await another stimulus.

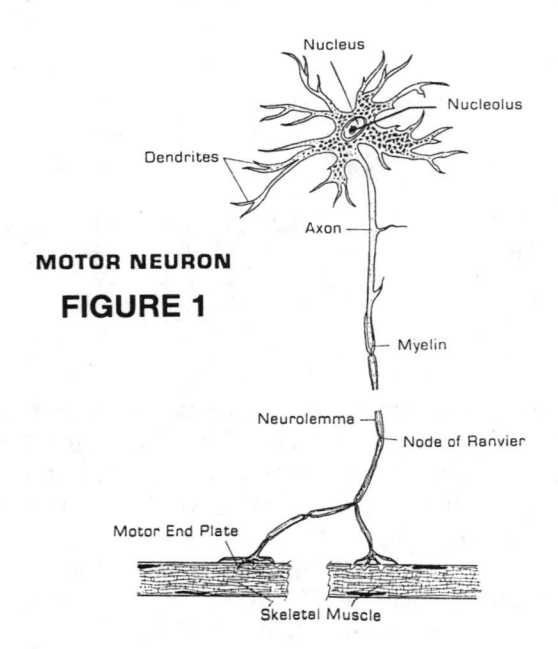

MOTOR NEURON
FIGURE 1

Neurons may contain a lipid coat called a **myelin sheath**. This sheath, if present, is interrupted at intervals by gaps in the sheath called the **nodes of Ranvier**. Myelinated neurons are seen in the white matter of the CNS and in the PNS. Nonmyelinated neurons and nerve cell bodies make up the gray matter of the CNS. Myelin is covered by a sheath called the **neurolemma**. The neurolemma is found only in the peripheral nervous system. Its function is the regeneration of nerve fibers. Nerve tissue of the CNS and optic and auditory nerves cannot repair themselves as they lack a neurolemma.

The central nervous system is composed of the brain and spinal cord. The brain and spinal cord consist of gray matter and white matter. **Gray matter** is responsible for the passage of impulses across synapses while **white matter** is concerned with the conduction of impulses along nerve fibers. Spinal cord nerve fibers make up **nerve tracts**. A tract of nerve fibers has the same beginning, end, and function. Ascending tracts carry impulses toward the brain; descending tracts carry impulses away from the brain.

The brain is separated into several parts. The largest part of the brain is the **cerebrum**. The cerebrum is responsible for conscious and subconscious thought, memory, and intelligence. Connecting the cerebrum to the midbrain are the thalamus and hypothalamus. The **thalamus** acts as a relay station for sensory nerve impulses, and is involved in complex reflex motion and emotions. The **hypothalamus** also serves as a relay center, as well as controls temperature, appetite, thirst, sleep, and digestion. The **midbrain** is the center for pupillary reflexes and eye movement. The **pons** serves as a bridge between the midbrain and medulla. Its function is to transport messages. The pons and medulla make up the brain stem. The **medulla** controls heart rate, respiration, and blood pressure, as well as swallowing, vomiting, hiccoughing, coughing, and sneezing. The **cerebellum** coordinates muscular activity and regulates such unconscious activities as standing, grasping, walking, and so forth.

The **spinal cord** serves as a reflex center and pathway for ascending and descending tracts. It is made of gray matter surrounded by white matter. Reflex actions occur within the spinal cord. A **reflex action** is a motor response to a sensory message that does not require brain participation. As the response is made to the sensory message, the brain receives the message. The reflex arc responsible for responses without brain intervention consists of a receptor, a sensory neuron, occasionally an internuncial neuron, a motor neuron, and an effector (generally a muscle).

The brain and spinal cord are protected by the cranium (skull), the vertebrae, a cushion of cerebrospinal fluid and three layers of membranes termed meninges. The outermost meninx or membrane is the **dura mater**. The **arachnoid membrane** is the middle membrane, and the innermost membrane is the **pia mater**. The subarachnoid space, between the arachnoid and pia mater is filled with cerebrospinal fluid. The meninges protect the brain and spinal cord against injury and disease–causing organisms.

LABORATORY EXERCISE PROCEDURE

Carefully complete each exercise and the lab report sheet. When completed, submit the lab report sheets to the instructor for grading.

LABORATORY EXERCISES

I. Neurons

   Study figure 127d, pg. 127 in the *Photo Atlas*. Sketch a neuron on the lab report sheet Section I and label dendrites, axon, nerve cell body, and nucleus.

II. Structure and Function of the Brain

Study the figures of the sheep brain and the accompanying structure and function list. The report sheet has illustrations of the sheep brain. Label the structures listed below, keeping in mind that some structures will be visible on both views of the brain and should be labeled each time they appear if so indicated by a label line. Use **only** the number of the structure on the report sheet. Refer to page 138 in the *Photo Atlas* for help.

1. Central sulcus
2. Cerebellum
3. Cerebrum
4. Corpus callosum
5. Frontal lobe
6. Hypothalamus
7. Medulla Oblongata
8. Midbrain
9. ccipital lobe
10. Olfactory lobe
11. Optic chiasma
12. Parietal lobe
13. Pituitary
14. Pons
15. Spinal cord
16. Temporal lobe
17. Thalamus
18. Ventricle

Brain Structure and Function List

Dorsal view of the brain:

1. Cerebrum—Two cerebral hemispheres divided by a longitudinal fissure and covered with gyri (ridges), sulci and fissures comprise the cerebrum which functions to govern muscular movement, interpret sensory impulses and control emotional and intellectual processes.

2. Frontal lobe—Each hemisphere of the cerebrum is divided into four lobes by sulci or fissures. The frontal lobes are anterior to the central sulci.

3. Parietal lobe—Separated from the frontal lobe by the central sulcus and extending posteriorly to the parietooccipital sulcus.

4. Occipital lobe—Located posteriorly to the parietal lobe, the occipital is the smallest of the lobes of the cerebrum.

5. Cerebellum—A prominent fissure, the transverse fissure, separates the cerebrum from the cerebellum. The cerebellum is the second largest portion of the brain and produces unconscious movements in the skeletal muscles resulting in coordinated movement, maintains normal muscle tone and body equilibrium. Two lateral lobes and a central smaller lobe, the vermis, comprise the cerebellum.

6. Medulla oblongata—A continuation of the spinal cord which forms the most inferior portion of the brain stem, the medulla oblongata contains many cell bodies of cranial nerves.

Ventral view of the brain:

7. Temporal lobe—The lateral cerebral sulcus (fissure of Sylvius) separates the frontal lobe from the temporal lobe.

8. Midbrain—Four rounded structures, the corpora quadrigemina, compose the mid–brain which functions in auditory and visual reflexes.

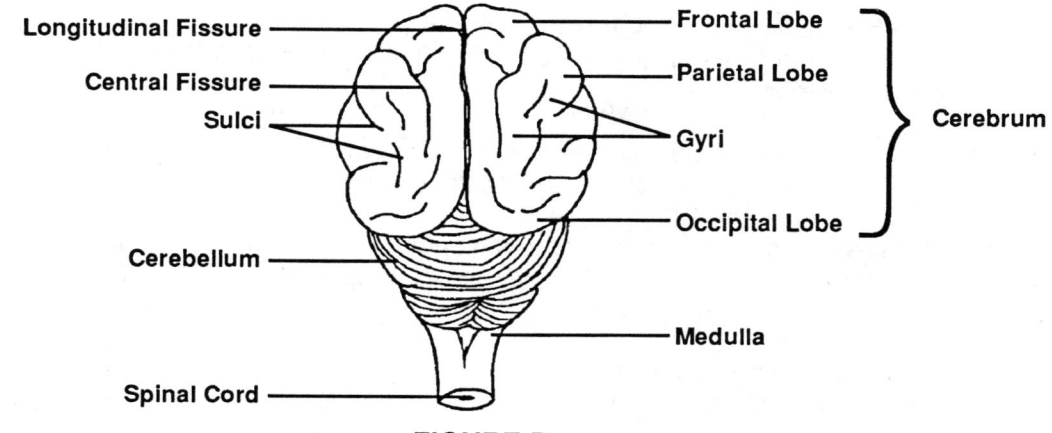

FIGURE B
Dorsal View of Sheep Brain

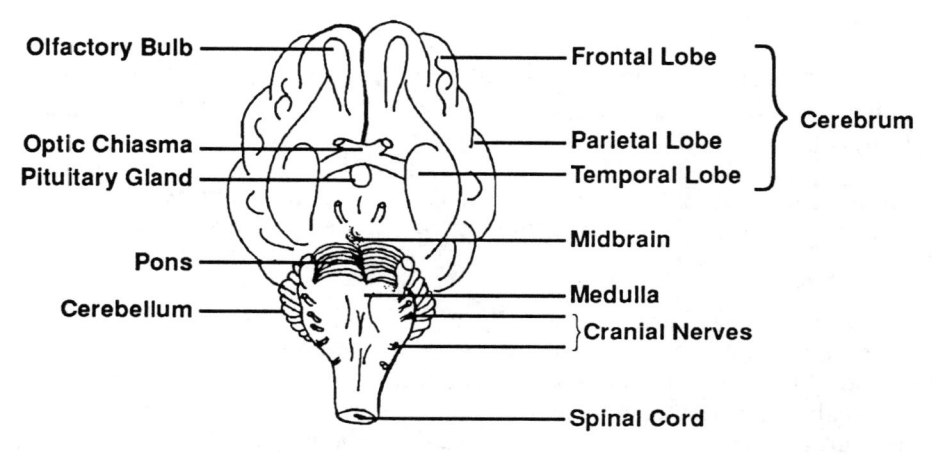

Labels: Longitudinal Fissure, Central Fissure, Sulci, Cerebellum, Spinal Cord, Frontal Lobe, Parietal Lobe, Gyri, Occipital Lobe, Medulla, Cerebrum

FIGURE C
Ventral View of Sheep Brain

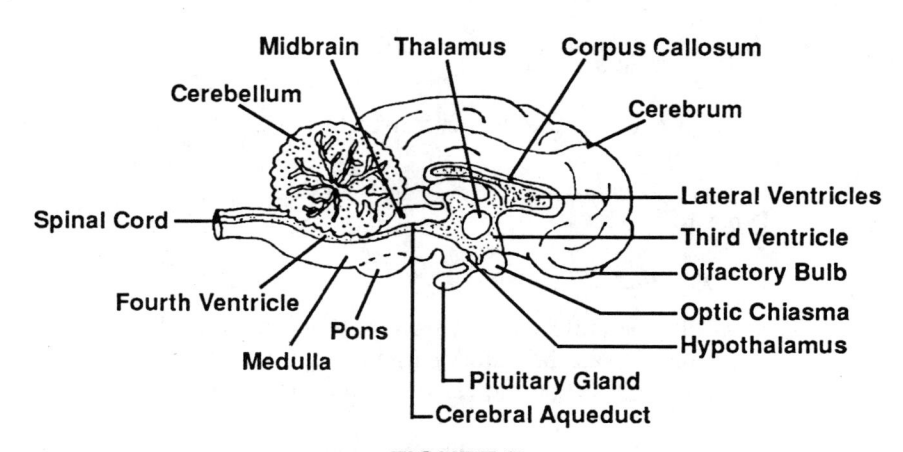

Labels: Olfactory Bulb, Optic Chiasma, Pituitary Gland, Pons, Cerebellum, Frontal Lobe, Parietal Lobe, Temporal Lobe, Midbrain, Medulla, Cranial Nerves, Spinal Cord, Cerebrum

FIGURE D
Sagittal View of Sheep Brain

Labels: Midbrain, Thalamus, Corpus Callosum, Cerebellum, Cerebrum, Spinal Cord, Lateral Ventricles, Third Ventricle, Olfactory Bulb, Optic Chiasma, Hypothalamus, Fourth Ventricle, Pons, Medulla, Pituitary Gland, Cerebral Aqueduct

FIGURE 2

The Nervous System 275

9. Olfactory bulbs—Located at the anterior of each cerebral hemisphere lies an olfactory bulb. The olfactory tracts lead posteriorly from the olfactory bulbs and carry sensations of smell.

10. Optic chiasma—Two transverse bands which cross in the median area of the ventral surface are the optic nerves leading from the eyes.

11. Pituitary gland—Lying posterior to the optic chiasma and ventrally to the midbrain, the pituitary gland is sometimes referred to as the master gland because it regulates so many body activities.

12. Pons—Anterior to the cerebellum and directly above the medulla oblongata, the pons connects the brain to the spinal cord and other parts of the brain to each other.

Sagittal section of the brain:

13. Corpus callosum—Under the cerebrum a thick, fibrous, white tissue which connects the cerebral hemispheres.

14. Thalamus—A large oval structure located under the corpus callosum and above the midbrain, the thalamus sends all sensory impulses responsible for the recognition of the sensation of pain.

15. Hypothalamus—Located ventral to the thalamus, the hypothalamus regulates body temperature, nutrient intake, sleep and waking states, and controls the autonomic nervous system.

16. Ventricles—Cerebrospinal fluid circulates through the ventricles of the brain as well as around the brain. The two lateral ventricles are located one in each cerebral hemisphere ventral to the corpus callosum. The third ventricle lies between the right and left halves of the thalamus and the lateral ventricles. It is a slit-like cavity. The fourth ventricle is located between the cerebellum and the medulla and pons. The cerebro-spinal fluid is produced in the ventricles and circulates around the brain and spinal cord carrying nutrients and serving as a shock absorber.

III. Reflex Experiments

The following experiments demonstrate reflexes. Transfer the results to the report sheet, Section III-A.

A. Types of reflexes

1. *Patella reflex or knee jerk*

    Sit on a table with legs crossed and hanging free. Relax the crossed leg. Sharply tap the patellar ligament just below the knee, using the side of the hand. Describe the result on the lab report sheet.

    Repeat the experiment while clinching a fist tightly just as the blow is about to be struck. Describe what happens on the lab report sheet.

2. *Plantar reflex*

   Kneel on a chair with both feet extending out over the edges of the chair. Have someone stroke the sole of a bare foot with a blunt object. Describe the results.

3. *Corneal reflex*

   Approach the cornea of the eye from the side, using a clean thread. Describe the results.

4. Observe the pupil size of the eye. Then cover it for a few seconds so that no light can get to it. Remove the cover and quickly note any change. Describe the results.

5. *Consensual reflex*

   Observe the pupils of both eyes. Then cover the right eye and note dilation of the pupil in the left eye. Uncover the eye and note any changes in the other eye.

6. *Accommodation reflex*

   With an assistant, observe the diameter of the assistant's pupil as he focuses his ttention on an object held at arm's length. Slowly bring the object closer to the eye and observe any change in the diameter of the pupil.

B. Reaction Time Experiment

   1. Objective: Determine reaction time by measuring how long it takes to catch a falling ruler.

   2. Materials: a ruler

   3. Procedure:

      a. Have an assistant hold a ruler by the 12 inch mark on it. Place your thumb and forefinger close to but not against the end of the ruler.

      b. When the ruler is released, try to catch it as quickly as possible, using only your thumb and forefinger.

      c. Record the number of inches the ruler fell, as measured by where your finger caught the ruler, in the chart.

      d. Repeat the experiment 4 times.

e. Using the information below, record the reaction time. Then total the time to get an average reaction time.

Inches Ruler Fell	=	Time in Seconds
1 inch		0.7
2 inches		0.10
3 inches		0.12
4 inches		0.14
5 inches		0.16
6 inches		0.17
7 inches		0.19
8 inches		0.20
9 inches		0.21
10 inches		0.22
11 inches		0.23
12 inches		0.24

f. Repeat the experiment again while being distracted. The assistant should ask you questions or have you do a math problem in your head as a distraction.

g. Determine your reaction time while distracted.

h. Write your results in the table in the report sheet.

IV. Using the *Photo Atlas*, draw a sketch of a Pacinian receptor and answer the questions in the lab report.

NERVOUS SYSTEM

Report Sheet 1

Name _____

Student ID # _____

Campus _____

Date _____

I.   Neuron—draw and label a neuron

Nervous System
Report Sheet 2

Student ID # _____

II. Brain Structure

A. Label the drawings below. These answers will be numbers.

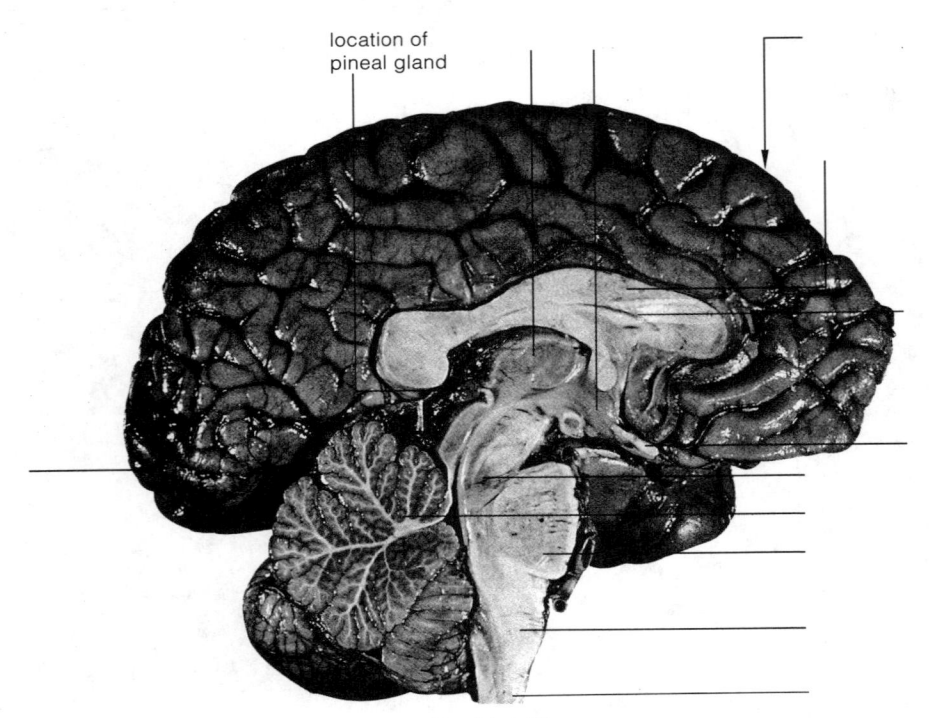

location of pineal gland

Sagittal View

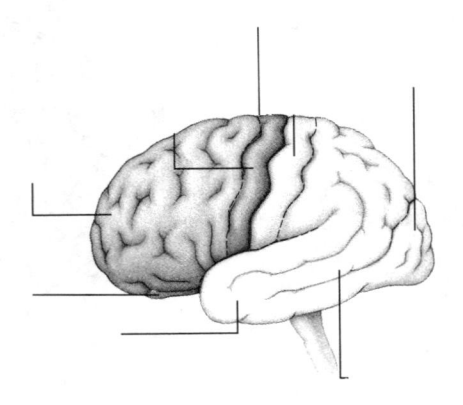

FIGURE 3

280   Exercise 26

Nervous System Student ID # _____
Report Sheet 3

    B. List, in order of their occurrence from the brain surface outward, the meninges of the brain.

        1. _____

        2. _____

        3. _____

III. Reflex Experiments

    A. Types of reflexes—Fill in the following blanks:

        1. Patellar reflex or knee jerk.

           Result? _____

           What happens with a clinched fist? _____

           _____

           Explain_____

           _____

        2. Plantar reflex

           Result? _____

        3. Corneal reflex

           What happens? _____

        4. Light reflex

           Result? _____

           _____

        5. Consensual reflex

           Result? _____

           _____

Nervous System
Report Sheet 4

Student ID # _____

6. Accommodation reflex

   Result? _____

   _____

B. Reaction Time Chart

| | Normal Conditions || Distracted Conditions ||
Trial	Inches Ruler Fell	Time in Seconds	Inches Ruler Fell	Time in Seconds
1				
2				
3				
4				
5				

Total = _____   Total = _____

Average = _____   Average = _____

C. Answer the following questions concerning the reaction time experiment.

1. What was the average reaction time when not distracted? _____

   When distracted? _____

   Why? _____

   _____

2. Would a reaction time of zero be unusual? _____ Why or why not?

   _____

   _____

3. Why would distractions while driving an auto or riding a motorcycle be dangerous?

   _____

   _____

Nervous System
Report Sheet 4

Student ID # _____

IV. The Pacinian Receptor

   A. Figure 128d, pg. 128 of the *Photo Atlas* shows a Pacinian receptor in cross section.

   The _____ in the center is surrounded by _____.

   The tiny dark dots are cell nuclei.

   B. A Pacinian receptor is a _____ receptor found in the _____ .

27 SENSORY PERCEPTION

LESSON OBJECTIVES

Upon completion of this laboratory exercise the student will be able to:

1. Identify the three classifications of sensory receptors.

2. Conduct experiments designed to investigate the ability of humans to hear and see.

3. Identify the structure and function of the parts of the ear and eye.

4. Learn the mechanisms responsible for different types of sensory perception.

5. Conduct experiments designed to investigate the ability of humans to smell, taste, and respond to touch, heat and cold.

MATERIALS NEEDED

1. Laboratory Manual
2. Pen or pencil
3. Mirror (hand or wall type)
4. Cotton–type applicators or swabs such as Q–tips
5. Sugar solution (1/4 teaspoon sugar dissolved in 1/4 cup water)
6. Granulated sugar
7. Salt grains
8. Salt solution (1/4 teaspoon salt dissolved in 1/4 cup water)
9. Lemon juice
10. Paper towels
11. Epsom salt solution
12. Water maintained at $0°—5°$ C ($32°—41°$ F)
13. Water maintained at $55°—60°$ C ($131°—140°$ F)
14. Two stainless steel teaspoons
15. Hairpin
16. metric ruler
17. Cider vinegar
18. 2 thermometers (one should be a cooking or candy thermometer for use in the hot water; one will be used in cold water)
19. Clock or watch with a second hand

PREPARATION

Read the discussion and the appropriate chapter(s) in your textbook before attempting to complete this lab exercise. Follow the directions in the laboratory manual carefully.

DISCUSSION

The ability to perceive conditions in the surrounding environment varies from individual to individual. We are constantly presented with forms of energy which require interpretation by our receptors, the sense organs. More specifically, the sensory receptor cell is the point of contact between the nervous system of an organism and its environment. As the receptor cells are stimulated, nerve impulses are generated which can be interpreted by the nervous system as to intensity and duration.

The interpretation, in a specialized area of the cerebral cortex, of an impulse arising from receptors is called **sensation**. The receptors report changes taking place either within or outside the body. Sense organ receptors, a pathway to the brain, and sensory areas of the brain make up a sensory mechanism. Sensory receptors are the dendrite endings of afferent neurons located in the sense organ. The function of sensory receptors is to receive messages.

The brain serves as the interpreter and dictates by way of the nervous system what the individual's response will be. Thus, we have a pathway over which stimuli messages are transported and responded to: namely a sensory pathway consisting of a stimulus, a sensory cell or receptor, the brain and reaction centers where appropriate responses occur.

Sensory receptors can be classified as: external receptors or receptors affected by changes in the external environment, proprioceptors or receptors related to balance, and internal receptors or receptors affected by changes in the internal environment or viscera.

The receptors for touch, cutaneous pain, pressure, heat, cold, smell, taste, vision, and hearing are **external receptors. Proprioceptive receptors** are located in various skeletal muscles, tendons and joints, and in the semicircular canals. **Internal receptors** are visceral pain receptors, hunger receptors and thirst receptors.

Sensory receptors are also classified as to sensitivity to stimulus quality and include mechanoreceptors, ranging from touch to auditory receptors; chemoreceptors, olfactory and taste receptors; and photoreceptors, which respond to the visible portion of the electromagnetic spectrum.

Excitation of photoreceptors is responsible for visual perception. The accessory structures associated with the eye are the eyelashes, eyelids, and eyebrows. These structures provide protection against injury and invasion of the eye by foreign particles. The **eyelids**, composed of muscular tissues with a layer of skin on the outside and a specialized epithelial lining—the **conjunctiva**—on the inside, glide smoothly over the eyeball because the surface is constantly moistened by fluid secreted by the lacrimal glands. The **lacrimal glands** are located at the outer corner of each eye. As the fluid is released from the gland, it flows over the eyeball toward the inner corner where it collects and drains into the nasal passage.

The color portion of the eye is the iris. The **iris** is a circular, thin diaphragm containing pigment. The diaphragm adjusts to control the amount of light entering into the eye. The dark spot in the center of the iris is actually the opening through which light enters and is known as the **pupil.**

Surrounding the iris is the white portion of the eye known as the **sclera** which is an extremely tough membrane. The sclera becomes transparent toward the front of the eye to form the **cornea**. The sclera functions to protect and help maintain the rigidity of the eyeball. Between the cornea and the lens is a space filled with a watery substance termed **aqueous humor**. This is a transparent fluid which helps maintain the shape and pressure of the eye and through which light can pass.

Six muscles account for eye movement from side to side and up and down. On the back portion of the eye is a projection. This projection is the **optic nerve** which extends from the eye to the brain. It is a large, tough nerve cord which transmits sensory impulses associated with visual perception.

The eye has two cavities: the anterior cavity or space in front of the lens and the posterior cavity or space posterior to the lens. The anterior cavity is divided into the **anterior chamber** or space between the cornea and the iris, and the **posterior chamber** or space between the iris and the lens. Both spaces are filled with aqueous humor which maintains intra-ocular pressure and the shape of the eyeball. The posterior cavity is filled with **vitreous humor** which serves as a transparent medium through which light passes and helps maintain the internal pressure and shape of the eye.

The **lens** is a transparent biconcave structure composed of fibrous protein which has been laid down in concentric lamellae (layers). The lens is held in place by **ciliary muscles** whose function is to change the shape of the lens which results in proper focusing. The lens is elastic and thus the shape can be altered somewhat.

Three layers form the wall of the eyeball. The outer layer is the sclera previously discussed. The middle layer is the **choroid layer**. It is a black or blue opaque structure with numerous blood vessels. The choroid layer functions to absorb excess light.

The innermost layer is an extension of the optic nerve termed the retina. In nature the retina is a thin structure consisting of an outer sensory layer containing the photoreceptors (rods and cones) and an inner layer which organizes and relays impulses generated in the photoreceptor layer to the brain. Two areas of interest found on the retina include the **fovea centralis** which lies in the optical axis of the eye, contains only cones, and is the region of most acute vision, and the **blind spot**. The area on the retina where all the retinal nerve fibers converge to exit from the eye as the optic nerve is known as the blind spot due to the absence of rods or cones in this area.

Rods are used for night or dim light vision, and **cones** are used for daytime vision. Stimuli in the form of impulses picked up by rods and cones is transmitted to the optic nerve for transmission to the brain.

Auditory sensations and equilibrium are the result of proper functioning of the ear. The ear can be subdivided into three principal regions: the outer ear, the middle ear and the inner ear.

The **outer** or **external ear** functions to collect sound waves and direct them inward. The visible portion of the outer ear is the **pinna** which is a skin-covered flap of elastic cartilage. Leading from the pinna to the middle ear is the **auditory canal** through which sound travels inward toward the **tympanic membrane** or **eardrum**. The auditory canal contains a few hairs plus ceruminous glands which secrete **cerumen** or earwax. The function of the hairs and wax is protection and lubrication.

The **middle ear** is separated from the outer ear by the tympanic membrane and from the inner ear by a thin bony structure which contains two openings—the round and oval windows. Located in this small chamber known as the middle ear are three small bones known collectively as the auditory ossicles. These three tiny bones are named the **malleus** (hammer), **incus** (anvil) and **stapes** (stirrup); the names being descriptive of their individual shapes.

Sensory Perception

The malleus is attached to the inner portion of the tympanic membrane. As sound waves strike the tympanic membrane, the malleus is set in motion. The head of the malleus sets in motion the anvil which in turn causes movement of the stapes against the oval window otherwise called the **fenestra vestibuli**. The attachment of the auditory ossicles to the oval window, the tympanic membrane, and to each other is by means of muscles and ligaments.

There is an opening on the anterior wall of the middle ear leading to the pharynx. This is the **Eustachian** or **auditory tube** which serves to equalize pressure on both sides of the tympanic membrane. Thus, deliberate swallowing may serve to equalize sudden pressure changes on the eardrum.

The inner ear is called the **labyrinth** because it is composed of a series of complicated, interconnected canals. The structure concerned with hearing and resembling a snail's shell is called the **cochlea**. Internally the cochlea consists of a bony spiral which makes two and one-half turns around a central bony core. This canal is divided into three fluid filled chambers, the **scala tympani**, the **scala vestibuli** and the **cochlea duct**. The **basilar membrane** separates the scala tympani from the cochlea duct. Located on the basilar membrane are the receptors known as the **Organ of Corti** which convert sound waves into nerve impulses. Associated with the cochlea is the cochlear branch of Cranial Nerve VIII, the vestibulocochlear or auditory nerve. As sound waves are converted into nerve impulses in the Organ of Corti they are passed on to the medulla by the auditory nerve.

The three **semicircular canals** along with two small sacs—the **saccule** and the **utricle**—are responsible for equilibrium. The three semicircular canals are at right angles to each other. The semicircular canals contain hair cells and a fluid called **endolymph** which flows over the hair cells. As the head is turned, the fluid moves setting the hair cells in motion. The movement of the hair cells stimulates the sensory neurons and the impulses are transported to the temporal lobe of the cerebrum by the cochlear portion of the vestibulocochlear nerve. The brain sends impulses to the proper muscles which must contract to maintain body balance.

The semicircular canals are responsible for **dynamic equilibrium**. The saccule and utricle are concerned with **static equilibrium** which is the orientation of the body relative to the surface of the earth. These hollow sacs are lined with sensory hair cells which are coated with a gelatinous substance containing **otoliths** (calcium carbonate particles). As the head is moved, the otoliths adjust in relation to the earth's gravitational field. The movement of the otoliths moves the hair cells which stimulates the receptor cells at their bases. The impulses are transported to the temporal lobe of the brain through the auditory nerve.

LABORATORY EXERCISE PROCEDURE

Carefully complete each exercise indicated and record the required information on the report sheets. When complete, submit the report sheets to the instructor for grading.

LABORATORY EXERCISES

Assemble all of the needed materials and supplies listed in the Materials section before proceeding. Follow the procedure and directions very carefully to ensure good results.

I. The Eye—A drawing of the eye is on Report Sheet 1. Label the indicated structures and give the function of each structure.

II. The following exercises are simple tests that reveal both normal and abnormal characteristics of the eye.

A. The Blind Spot

The eyes can be tested to locate the blind spot by using the chart following. To test the right eye, close the left eye and stare at the plus sign while moving the paper from 18 inches away toward the face. At first both the plus and the dot will be seen. At some point the dot will no longer be visible as it comes into focus on the blind spot of the retina. Hold the paper at the spot where the dot disappeared and measure its distance from the eye. Record this measurement in the lab report. Now repeat the experiment with the left eye. Again record the results in the lab report.

B. Near Point Accommodation

The closest distance at which the eye can see a clear undisturbed image is the near point. The older a person becomes, the more difficult it is to focus due to a loss in elasticity of the lens. The normal near point at age 20 is 3-1/2 inches; age 30, 4-1/2 inches; age 40, around 6-3/4 inches. By age 50 it is 20-1/2 inches.

Use the letter U at the beginning of this sentence to help determine the near point in both eyes together and in each eye. Move this page toward the face until the letter becomes blurred. Then close one eye and move this page toward the eye until the letter becomes blurred. Next move the page back until the image is clear. Measure this distance from the eye to the page. Test the other eye the same way and record the results in the lab report.

C. Light Accommodation

To demonstrate pupil reaction to changes in amount of available light, cover the left eye with the left hand while looking into a mirror. After approximately one minute remove the left hand and quickly observe pupil sizes. Answer the questions on the report sheet, Section II C.

D. Astigmatism

Astigmatism is the result of the lens in an eye having different curvatures in different axes. When the lens has part of an object in focus in one axis, the object is blurred in another axis. Look at the astigmatism test chart and observe if all radiating lines are in focus and have the same intensity of blackness. No astigmatism exists if all the lines look the same. **Caution**: The presence of other refractive abnormalities may make this test impractical. Test each eye by closing the other and looking at the chart. Record the results in the lab report.

E. Illusions Perceived by the Eye

1. Look at the drawings of the three people. Which is the tallest? Measure the height of each of the people. Now which is the tallest? Many times the message about what is seen is incorrect. Backgrounds affect what you think you see by creating an illusion. Answer the questions in the lab report.

2. Stare at the cube for a few seconds. Does the cube appear to reverse itself? It should appear to have reversed itself with the shaded section changing from being a top to being a side of the cube.

Sensory Perception **289**

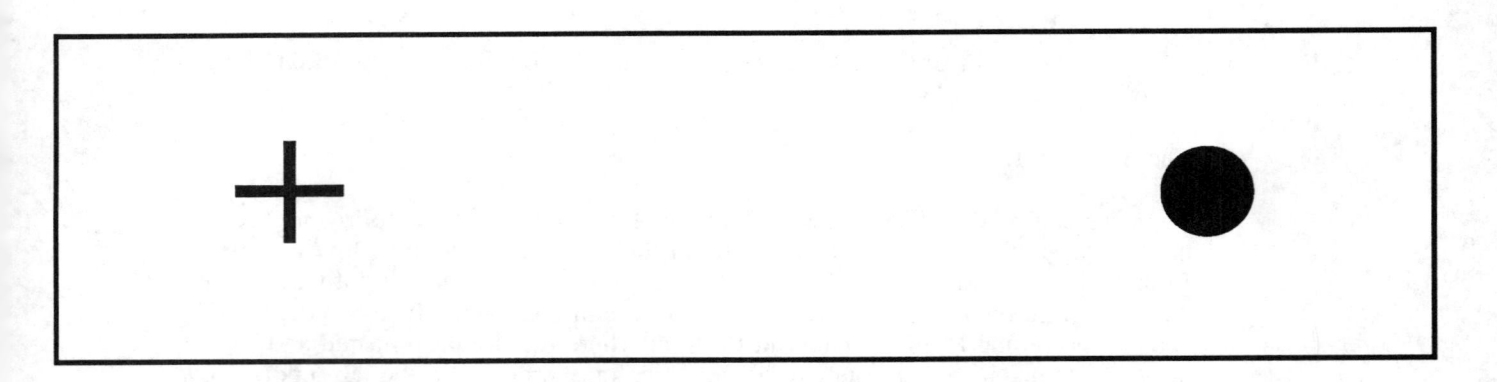

BLIND SPOT TEST

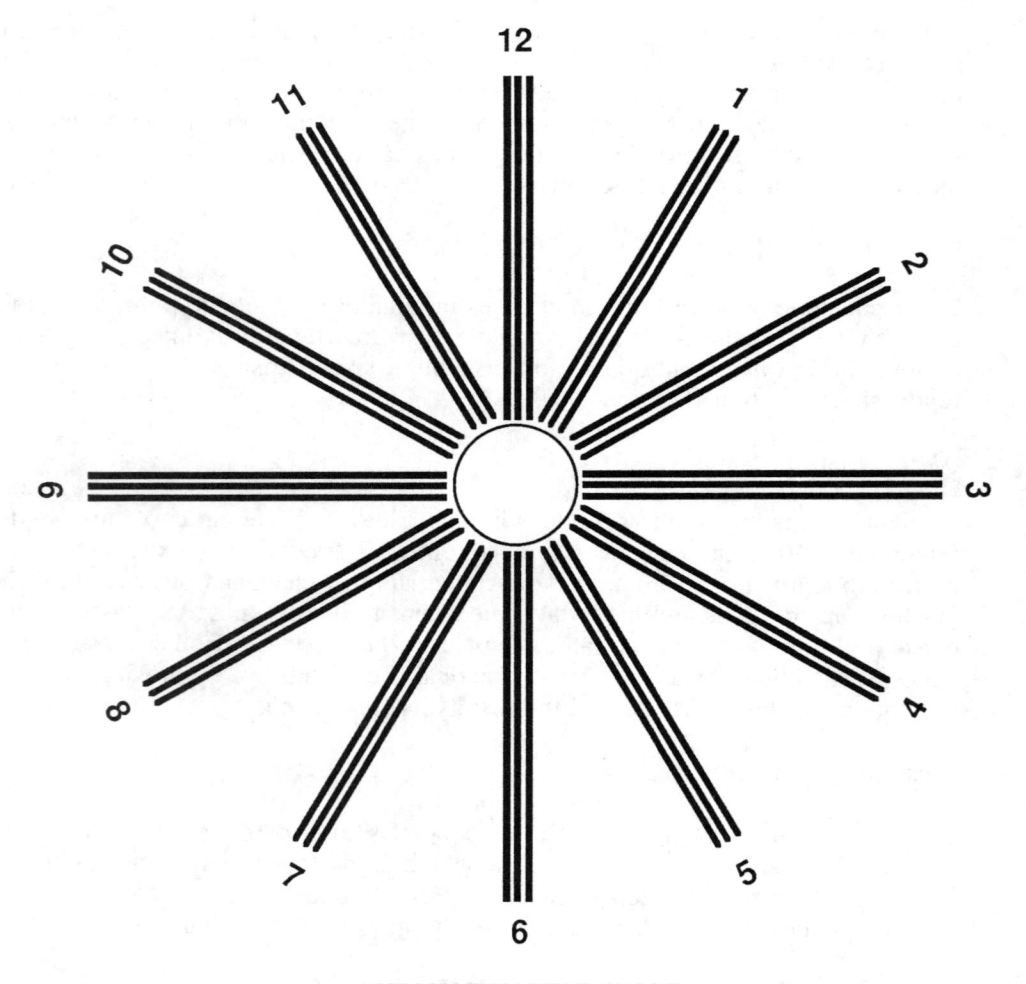

ASTIGMATISM TEST CHART

FIGURE 1

290   Exercise 27

FIGURE 2

III. The Ear—Locate the pinna; ear drum; earbones (malleus, incus, stapes); auditory nerve; cochlea; and semicircular canals on the drawing of the ear and give the function of each structure.

IV. Perform the simple hearing experiments as described on the lab report sheet and answer the corresponding questions.

V. Open the container of cider vinegar and bring the container close to the nose (about 6 inches or 15 cm). The pungent applelike odor of cider vinegar was detected and received by the olfactory sensory cells which are located high in the nasal cavity roof. On the Report Sheet identify the stimulus, the location of the sensory cells involved, the interpreter and describe the response to the stimuli (Section V).

VI. The ability to distinguish between bitter, sour, sweet and salty tastes enhances meals and is responsible for food preferences and dislikes. The tongue contains taste buds—approximately 8–9000 on the human tongue. As food is dissolved in the mouth and broken down into chemical solution, the liquid stimulates the taste bud cells to send a message to the brain. The sense of taste is referred to as the process of gustation or the gustatory sense.

Use one of the clean paper towels to blot the tongue. Place a few grains of salt on the tip of the tongue. How long did it take to perceive a salty taste? Blot the tongue once again and repeat the procedure using a cotton–tipped applicator dipped in the salt solution. On the Report Sheet record the length of time required to perceive a salty taste using the salt solution. Once again blot the tongue. This time drop several granules of sugar on the tongue approximately one inch from the tip. Record the time required to perceive a sweet taste on the Report Sheet VI C. After blotting the tongue, repeat the procedure using a cotton–tipped applicator dipped in the sugar solution. Record the observation on the Report Sheet VI D.

The taste receptors responded quicker to which substances—the grains of salt and sugar or the solutions of salt and sugar? Record the answer on the Report Sheet VI E.

VII. Bitter and sour tastes are perceived by other areas of the tongue. Blot the tongue and test for these areas using separate cotton–tipped applicators for the sour (lemon) and bitter (epsom salt) solutions. The use of a mirror will be useful in this investigation. On the Report Sheet is a diagram of the tongue. Indicate the areas responsible for the perception of salty, sweet, sour and bitter tastes by using arrows, labels, and shading.

VIII. The receptors for the sensations of touch, pressure, cold, heat and pain are located in the skin connective tissue and the ends of the gastrointestinal tract. These are known as the cutaneous sensations. The cutaneous receptors are distributed over the body but not uniformly as some areas contain many more receptors than do other areas. A simple demonstration of this random distribution of receptors known as the two–point discrimination test for touch is easily accomplished. Using the metric ruler adjust the distance between the tips of the hairpin as indicated on the chart found on the Report Sheet, Number VIII. Touch the indicated area, record the observation, and repeat the procedure adjusting the distance between the tips of the hairpin as indicated on the chart. Complete Section VIII of the Report Sheet.

IX. The receptors for sensations of heat and cold are located in the dermis. As with the touch receptors, the heat and cold receptors are distributed randomly throughout the body. To investigate this and compare relative sensitivities to heat and cold stimuli, perform the following tasks and answer the questions on the Report Sheet, Section IX.

Prepare two containers of water; one with a thermometer in it which reads 0°–5° C (32°–41° F), and another in which the water is warm—550°–60°C (131°–140° F). The cooking or candy thermometer should be used in the warm water. The handles of the teaspoons will serve as probes.

Lay the metric ruler along the forearm, holding it in place with the tips of the fingers or with tape. (If right handed, place the ruler on the right forearm.) Holding the teaspoon by the bowl, dip the handle into the cold water, touching it to the skin of the forearm every 4 centimeters. Repeat this procedure using a teaspoon directly dipped in the hot water. Record the sensations on the Report Sheet, Section IX.

X. Generally, the brain interprets signals so that we understand what they mean. Sometimes, however, the brain may misinterpret signals sent to it by the sense organs. Set up three pans: one containing hot water, one containing water at room temperature, and one containing ice water. **CAUTION**: Extremely hot water will burn the hand! Place the left hand in the hot water (131°–140°F) and at the same time place the right hand in the cold water. Leave the hands in the water for 30 seconds. At the end of 30 seconds remove the hands from the hot and cold water and place both of them in the pan containing water at room temperature. How does the room temperature water feel to each hand? Record the answer on the Report Sheet, Section X.

Based on this experiment, can the sense of touch be fooled? Record the answer on the Report Sheet, Section XB. If so, under what conditions does the sense of touch seem to be fooled? Record the answer on the Report Sheet, Section X.

SENSORY PERCEPTION

Report Sheet 1

Name _____

Student ID # _____

Campus _____

Date _____

I. The Eye

Locate the following on the cross–section of the eye and give the function of each structure:

a. Retina   c. Cornea   e. Lens
b. Optic Nerve   d. Pupil   f. Iris

Structure   Function

These answers will be letters.

1. _____   _____

2. _____   _____

3. _____   _____

4. _____   _____

5. _____   _____

6. _____   _____

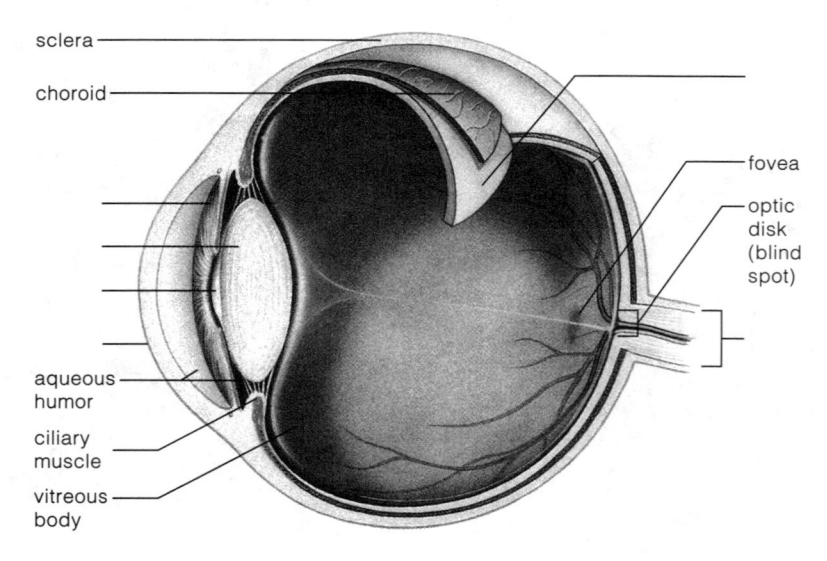

Structure of the human eye.

FIGURE 3

294   Exercise 27

Sensory Perception　　　　　　　　　Student ID # _____
Report Sheet 2

II. Eye Experiments

    A. Blind Sport

        1. What is the blind spot? _____

            _____

        2. Where is the blind spot located? _____

            _____

            _____

        3. Record the blind spot measurements

        Right eye _____

        Left eye _____

    B. Near Point Accommodation

        1. What is the near point? _____

            _____

            _____

        2. What can cause difficulty in focusing? _____

            _____

            _____

        3. Record the near point accommodation measurement.

        Both eyes _____

        Right eye _____

        Left eye _____

Sensory Perception
Report Sheet 3

Student ID # _____

C. Light Accommodation

1. What structure in the eye functions to protect the eye from changes in light intensity?

   _____

2. Describe what happened to the pupil when the hand was removed from the eye.

   _____

   _____

   _____

D. Astigmatism

1. What is astigmatism? _____

   _____

   _____

2. What causes astigmatism? _____

   _____

   _____

3. Record the astigmatism test results: Vision Blurred or Clear

   Right eye _____

   Left eye _____

E. Illusions

1. In the drawing of the three people, which one is tallest? Circle the correct answer.

   A   or   B   or   C

2. Which person measured the tallest?   A   or   B   or   C   (Circle the correct answer.) Explain the answer.

   _____

   _____

Sensory Perception   Student ID # _____
Report Sheet 4

    3. Did the cube reverse itself?_____

    4. Describe another type of illusion. (Not one mentioned in this lab.) _____

_____

_____

III. The Ear

Locate the following on the ear cross–section and give the function of each structure:

Pinna	Ear Bones (ossicles)	Auditory Nerve
Ear Drum	Semicircular Canals	Cochlea
Stirrup or stapes	Hammer or malleus	Anvil or incus

        Structure   Function

1. _____   _____

2. _____   _____

3. _____ a. _____ _____

                b. _____ _____

                c. _____ _____

4. _____   _____

5. _____   _____

6. _____   _____

IV. Ear Experiments

Answer the questions after performing the following:

A. Block the ears with the fingers or cotton so that no sound is heard. Then softly read this sentence out loud. Can you hear yourself? _____ How do you sound?

_____

_____

Sensory Perception

a Functional divisions of the human ear

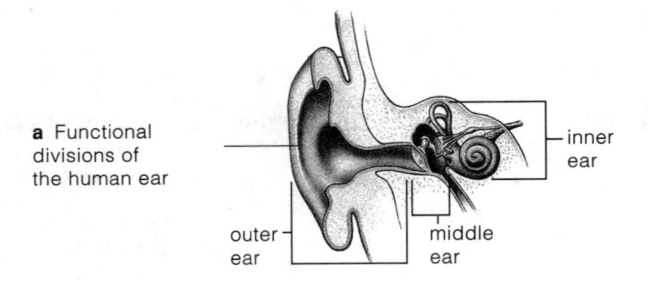

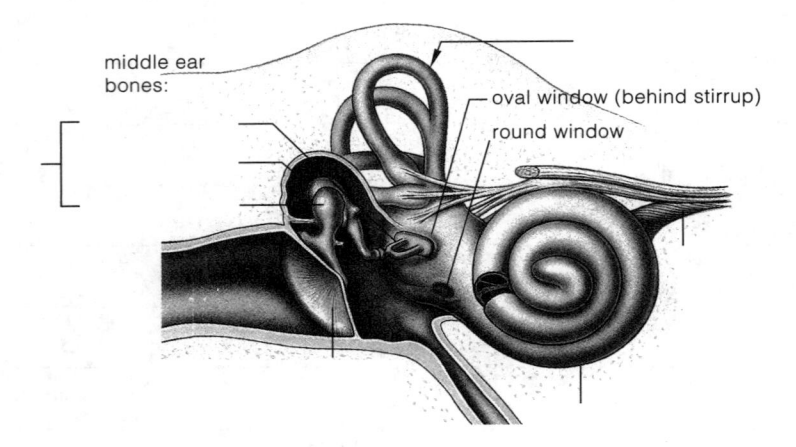

outer ear
middle ear
inner ear

middle ear bones:
oval window (behind stirrup)
round window

b Structure of the middle ear and the inner ear

FIGURE 4

Sensory Perception
Report Sheet 5

Student ID # _____

    B. Cup the hands behind the ears to help catch the sound. Now read this sentence out loud.

        Do you sound the same as before? _____ This is closer to the way you sound to others.

V. Olfactory Sensory Perception

    A. Stimulus _____

    B. Location _____

    C. Interpreter _____

    D. Response _____

        _____

VI. Taste Perception

    A. Approximate time to perceive a salty taste with salt grains _____

    B. Approximate time to perceive a salty taste with salt solution _____

    C. Approximate time to perceive a sweet taste with granulated sugar _____

    D. Approximate time to perceive a sweet taste with sugar solution _____

    E. _____

VII. Indicate the areas of the tongue responsible for the perception of salty, sweet, sour, and bitter tastes (by using arrows, labels, and shading).

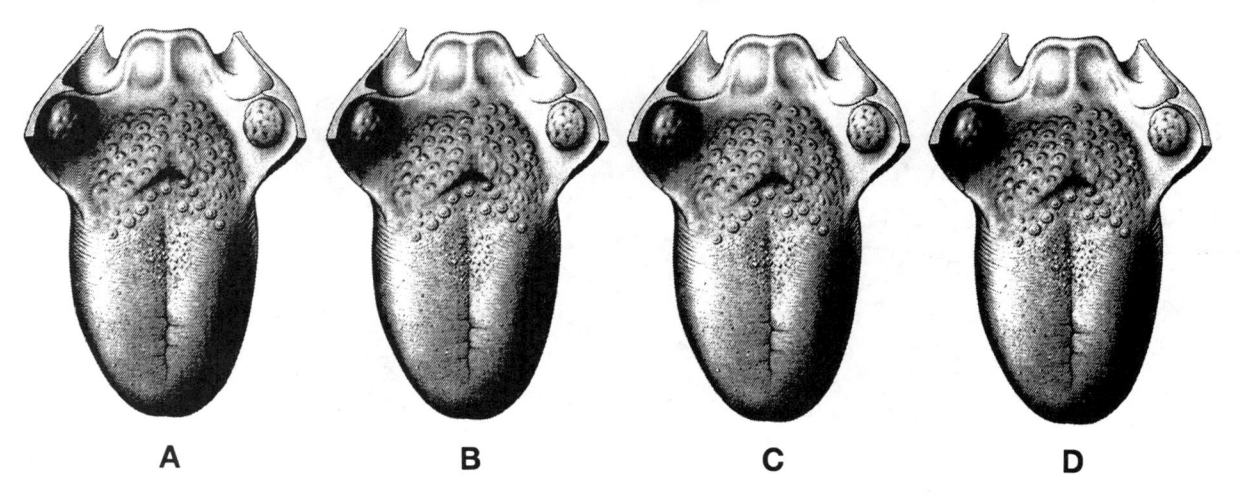

A   B   C   D

FIGURE 5

Sensory Perception
Report Sheet 6

Student ID # _____

VIII.

Area to Touch	Distance Between Hairpin Points	Felt as One or Two Points	Ranked as Most Sensitive (4) to Least Sensitive (1)
Back of hand	15.5 mm		
Tip of index finger	2 mm		
Tip of nose	3 mm		
Side of nose	3 mm		
Tip of tongue	1.4 mm		

IX. On the sketch of the forearm below, indicate areas which seemed most sensitive to the heat and cold. The most sensitive area to cold will be labeled C–1, the least sensitive C–4. The area most sensitive to heat will be labeled H–1, the least sensitive H–4. (If more than four areas were tested, adjust the number system accordingly.)

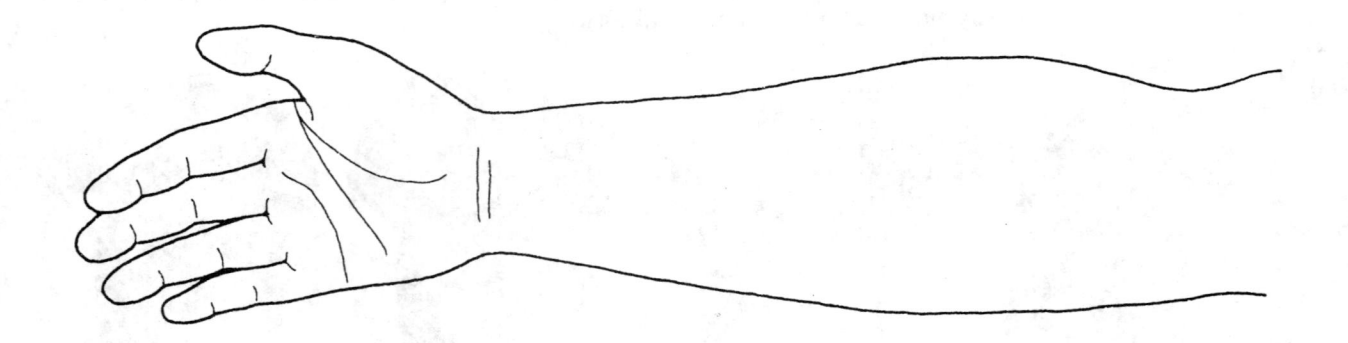

FIGURE 6

Sensory Perception Student ID # _____
Report Sheet 7

    A. Are the areas for cold perception the same as for heat? _____

    _____

    B. Does the ability to perceive heat and cold aid humans in their daily lives? How?

    _____

    _____

X. A. After immersion of one hand in cold water and the other in hot water, how does room temperature water feel to each hand?

    _____

    _____

    B. Can the sense of touch be fooled? _____

    C. If so, under what conditions does the sense of touch seem to be fooled? _____

    _____

    _____

28 REPRODUCTION

LESSON OBJECTIVES

Upon completion of this laboratory exercise the student will be able to:

1. Distinguish between asexual and sexual reproduction.

2. Distinguish between external and internal fertilization.

3. Trace the path of a sperm cell from the testes to outside the body.

4. Trace the path of an egg cell from the ovary to outside the body.

5. Discuss the steps in the menstrual cycle including all hormones in the discussion.

6. Be able to identify the structures of the male and female reproductive system and give their functions.

7. List three glands associated with the male reproductive system and describe the functions of their secretions.

MATERIALS NEEDED

1. Laboratory Manual
2. *Photo Atlas for Biology*

PREPARATION

Read the discussion carefully before attempting this exercise. Also read the appropriate chapter(s) in your textbook.

DISCUSSION

There are two types of reproduction: asexual and sexual. **Asexual reproduction** involves only one parent. No special organs are required. In asexual reproduction a single parent splits, buds, or fragments, giving rise to two or more offspring that are identical to the parent. Asexual reproduction occurs in both plants and animals. Primitive plants and animals frequently reproduce asexually, while more advanced species reproduce sexually.

Sexual reproduction involves two parents, each of which contributes one sex cell, or **gamete**, to a union which produces a new organism. The union of sex cells (gametes) is termed **fertilization**. Fertilization occurs in one of two ways. **External fertilization** involves the shedding of male and female gametes into some surrounding medium such as water. The sperm swim or are carried to the eggs by the current. In **internal fertilization**, the egg cells are retained within the female reproductive tract until after they are fertilized by sperm inserted into the female by the male.

Humans reproduce sexually as do most animals by internal fertilization. After fertilization, the egg develops into an embryo by undergoing a series of complicated changes and processes. The new organism shares the characteristics of both parents. The purpose of this lab is to acquaint you with the human male and female reproductive organs and how these function in the process of insuring the perpetuation of *Homo sapiens*.

The reproductive system of both the male and female consists of primary and accessory sex organs. In the male the paired **testes** or testicles are the primary organs of reproduction. The testes are contained within the **scrotum**, a pouch attached to the outside of the abdominal wall. The testes produce viable sperm only at a temperature lower than the rest of the body. If the testes do not enter the scrotum, but remain in the abdominal cavity, the person will be sterile; that is, he will not produce living **spermatozoa**.

Tough fibrous connective tissue termed the **tunica albuginea** envelops each testis and forms a network inward resulting in numerous (200–400) compartments or lobules. Each compartment contains **seminiferous tubules** which produce spermatozoa or sperm. Sperm formation is called **spermatogenesis** and takes about two days. **Interstitial cells**, also found within the testes, produce the male sex hormone **testosterone**.

The seminiferous tubules pass mature sperm from the testes to the **epididymis** which is a long (about 20 feet) coiled tube. Sperm are stored here until released during copulation. As the epididymis becomes less coiled, its walls thicken and it is referred to as the **vas deferens** or **ductus deferens**. Circular smooth muscle tissue is found in the walls of the vas deferens. This muscle tissue moves the sperm along by means of peristaltic motion.

The vas deferens enters the abdominal cavity by passing through the inguinal canal and extends over the top of the **urinary bladder**. On the posterior surface of the bladder the vas deferens joins the **ejaculatory duct** which propels the spermatozoa into the urethra. The **urethra** passes through the penis and empties to the outside. Urine from the bladder moves through the urethra during urination and semen passes through the urethra during sexual activity.

Semen is a mixture of secretions from the accessory glands plus sperm. It is slightly alkaline and acts as a buffer to neutralize the acidity of the female reproductive system. The accessory glands include the paired seminal vesicles, the prostate gland and the paired Cowper's glands. The **seminal vesicles** lie posterior to and at the base of the urinary bladder. The **prostate gland** secretes a thin white alkaline fluid into the urethra which contains nutrients for the active sperm. The paired **Cowper's glands** are located beneath the prostate on each side of the urethra. They are about the size and shape of peas and secrete a mucous–like alkaline fluid into the urethra. The semen serves as a medium for the transport of sperm, protects the sperm from acids in the female genital tract by its alkaline nature, lubricates the passages through which the sperm travel and provides nutrients for the active sperm.

The **penis** functions to introduce the sperm into the vagina of the female. It is composed of three cylindrical spongelike masses of erectile tissue, the innermost surrounding the urethra. The penis terminates in an enlarged region termed the **glans penis**. A flap of skin covers the glans penis at birth. This skin, the **prepuce** or foreskin, is usually removed surgically shortly after birth by a procedure termed circumcision.

Under the influence of sexual stimulation, the arteries entering the penis dilate while the veins constrict. Thus, more blood enters the penis than leaves resulting in an increase in size of the penis as it becomes firm and erect. The average volume of semen for each ejaculation is approximately three to four cubic centimeters containing about 100 million spermatozoa.

The female reproductive system is more complex than that of the male as it is designed to provide the nourishment and environment for a developing human. The **ovaries** are the primary reproductive organs in the female as the eggs or **ova** are produced here and are located in the abdomen on either side of the uterus. The ovaries have two functions: they produce egg cells and female sex hormones **estrogen** and **progesterone**. Each egg cell is enclosed in a sac called a **follicle**. As the egg cell matures, the follicle increases in size. A ripe follicle bulges out from the ovary. When ovulation occurs, the outer wall of the follicle bursts and the egg or ovum and fluid are released. The egg is released into the abdominal cavity where it is normally drawn up into the **oviduct** or **Fallopian tubes**. The oviducts partially surround the ovaries but are not attached to them. Wavelike motions of the fingerlike projections of the oviduct propel the released ovum into the oviduct. The cilia lining the oviduct, peristaltic contractions of the oviduct and the flow of tissue fluids combine to move the ovum along the tube. If sperm are present in the oviduct the egg may be fertilized. Usually only one egg is released at a time.

The oviducts empty into the **uterus** or womb. If the ovum has been fertilized it will implant itself in the endometrium lining of the uterine wall where it develops until the time of birth. An unfertilized egg generally will not reach the uterus before it disintegrates and is absorbed by the body.

The uterus connects at the **cervix** with a muscular tube called the **vagina** or birth canal. The vagina leads to the outside and functions as both the birth canal and the receptacle for the male penis during copulation.

The external structures of the female reproductive system are collectively termed the **vulva**. A fat pad located over the pubic bone is termed the **mons pubis**. Below the mons pubis lies the **clitoris**, a small erectile structure comparable to the male penis. Two pairs of lip-like folds surround the vaginal opening. The **labia minora** are the delicate inner folds and the **labia majora** are the outer folds. The labia minora surround the area termed the **vestibule**. The vestibule encircles the urethral orifice and the vaginal orifice (opening). A clear, lubricating fluid is secreted during sexual stimulation by the **glands of Bartholin** located on either side of the vestibule.

In both the male and female, puberty begins when the hypothalamus stimulates the anterior pituitary gland to release gonadotrophic hormones. These hormones are responsible for the maturation of sex cells, the production of the sex hormones, and the development of secondary sex characteristics in both the male and the female. Secondary sex characteristics in the male include beard growth, pubic hair growth, deepening of the voice due to a thickening and enlargement of the vocal folds, maturation of the seminal vesicles and prostate gland, and the development of larger and stronger muscles. Secondary sex characteristics in the female include growth of pubic hair, broadening of the pelvis, breast development, increase in size of the uterus and vagina, some voice quality changes, and the onset of the menstrual cycle.

Variations in the secretion of the gonadotrophic hormones in females leads to the **menstrual cycle**. A "typical" menstrual cycle is about 28 days. During the menstrual phase, the thickened endometrium of the uterus is sloughed off. During this time, **FSH** (follicle stimulating hormone) is released by the anterior pituitary gland to stimulate the development of a group of follicles in the ovary. Generally one follicle is mature about 12–14 days after it begins to develop. The growing follicle stimulates the production of **estrogen** which promotes the replacement of the uterine lining sloughed off during menstruation. The production of estrogen inhibits the further production of FSH and stimulates the production of **LH** (Luteinizing hormone). LH promotes ovulation. The cells of the ruptured follicle undergo a transformation to a **corpus luteum** which functions as an endocrine gland producing **progesterone**, the pregnancy hormone. Progesterone continues to build up the uterine lining. At this point the uterus is ready to receive a fertilized egg. If the egg is not fertilized, the corpus luteum regresses to a **corpus albicans** and the production of progesterone is reduced to zero. The uterine lining begins to slough off, resulting in the menstrual flow.

LABORATORY EXERCISE PROCEDURE

Read the directions for the exercise carefully before completing the exercise. When complete submit the lab report to the instructor for grading.

LABORATORY EXERCISES

I. Asexual Reproduction. Use the *Photo Atlas for Biology*, page 26.

After looking at fig. 26a, answer the appropriate questions on the lab report.

II. The Male Reproductive System

   A. Label the drawing of the male reproductive system on the report sheet using the Discussion as a guide. The terms needed are listed below. Use *only* the number of the term on the report sheet.

1. Cowper's Gland	5. Prepuce or foreskin	9. Testis
2. Ejaculatory Duct	6. Prostate Gland	10. Penis
3. Epididymis	7. Seminal vesicle	11. Urethra
4. Glans penis	8. Scrotum	12. Urinary Bladder

   B. Study fig. 143c on page 143 of the *Photo Atlas*. Sketch the seminiferous tubule.

   C. See the report sheet for this exercise.

   D. See the report sheet for this exercise.

   E. See the report sheet for this exercise.

III. The Female Reproductive System

   A. Label the drawing of the female reproductive system on the lab report using the Discussion as a guide. Use the terms listed below. The answers on the lab report sheet will be numbers.

1. Bladder	4. labium majora	7. Ovary
2. Cervix	5. labium minora	8. Uterus
3. Clitoris	6. Oviduct	9. agina

   B. Examine the ovary in figures 143b–f on page 143 of the *Photo Atlas*. In cross section, a follicle resembles a hollow ball with a developing oocyte in it. The egg is an oocyte until fertilization at which time it is technically an ovum. One or more corpus luteum (a follicle from which the oocyte has escaped) may be observed. These have thicker walls than the maturing follicles. Sketch the ovary.

C. See the report sheet for this exercise.

D. See the report sheet for this exercise.

E. See the report sheet for this exercise.

F. See the report sheet for this exercise.

REPRODUCTION

Report Sheet 1

Name _____

Student ID # _____

Campus _____

Date _____

I. Asexual Reproduction

   A. Sketch an example of budding. (fig. 26a, pg. 26)

   B. Answer the following questions: (Refer to your textbook for help.)

   1. Explain why buddy is a type of reproduction. _____
   _____
   _____

   2. Why is the formation of spores considered to be asexual reproduction?
   _____
   _____

   3. Can an organism reproduce both sexually and asexually? _____ If so, give an example.
   _____

Reproduction
Report Sheet 2

Student ID # _____

4. How does regeneration differ from binary fission? _____

_____

_____

5. Briefly describe vegetative propagation. _____

_____

_____

_____

_____

II. Male Reproductive System

    A. Label the following drawing of the male reproductive system. The answers will be numbers. (Refer to pages 119 and 121 in the *Photo Atlas* for help.)

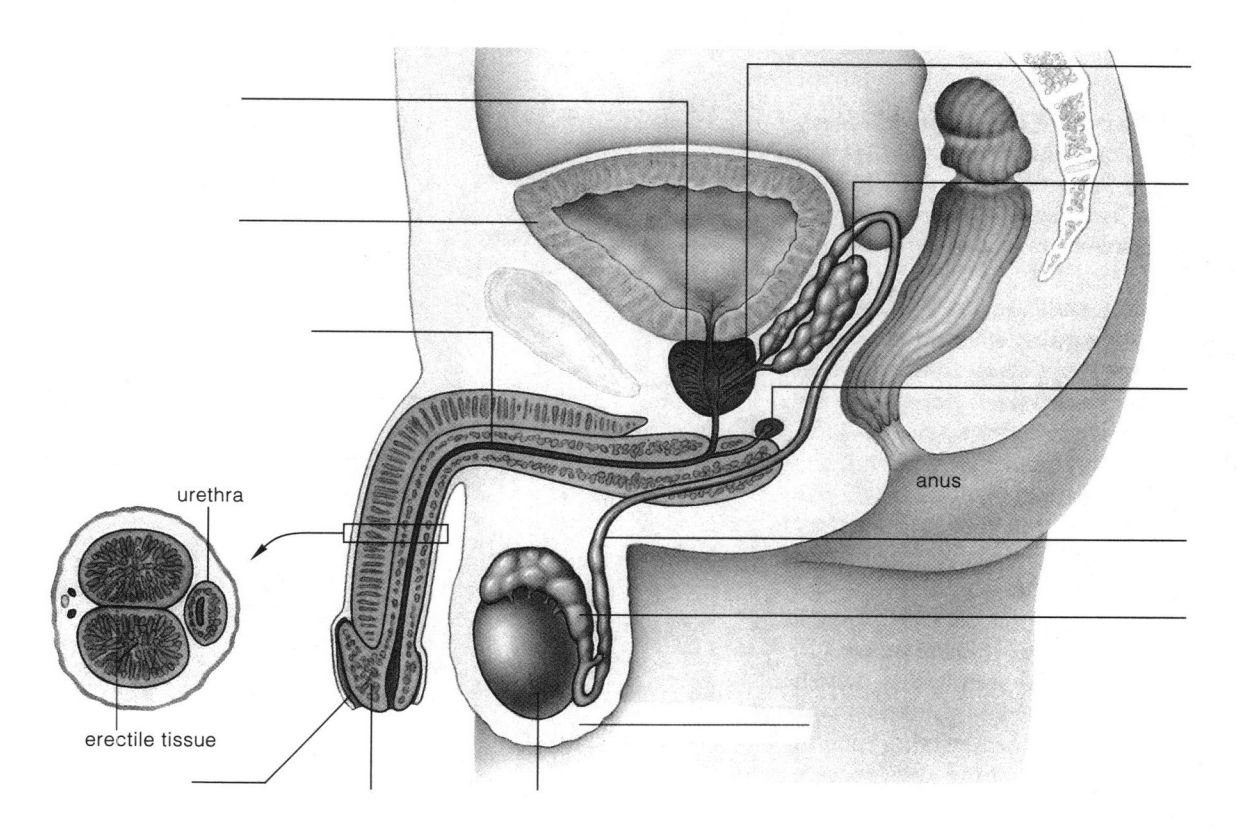

FIGURE 1

Reproduction **309**

Reproduction
Report Sheet 3

Student ID # _____

B. In the space provided below sketch a seminiferous tubule, labeling sperm, interstitial cells and seminiferous epithelium. (Fig. 143c, pg. 143)

C. Trace the path of a sperm cell from its origin in the testes to outside the body.

1. ____testis____   5. _____

2. _____   6. _____

3. _____   7. _____

4. _____

310 Exercise 28

Reproduction
Report Sheet 4

Student ID # _____

D. Give the function of the following structures:

    Structure   Function

1. Testes   _____

2. Scrotum   _____

3. Epididymis   _____

4. Vas deferens   _____

5. Seminal vesicle   _____

6. Prostate gland   _____

7. Urethra   _____

8. Cowper's gland   _____

9. Penis   _____

E. Semen consists of

1. _____

2. _____

3. _____

4. _____

Reproduction
Report Sheet 5

Student ID # _____

III. Female Reproductive System

   A. Label the drawing of the female reproductive system. The answers will be numbers. Refer to pages 119 and 121 in the *Photo Atlas* for help.

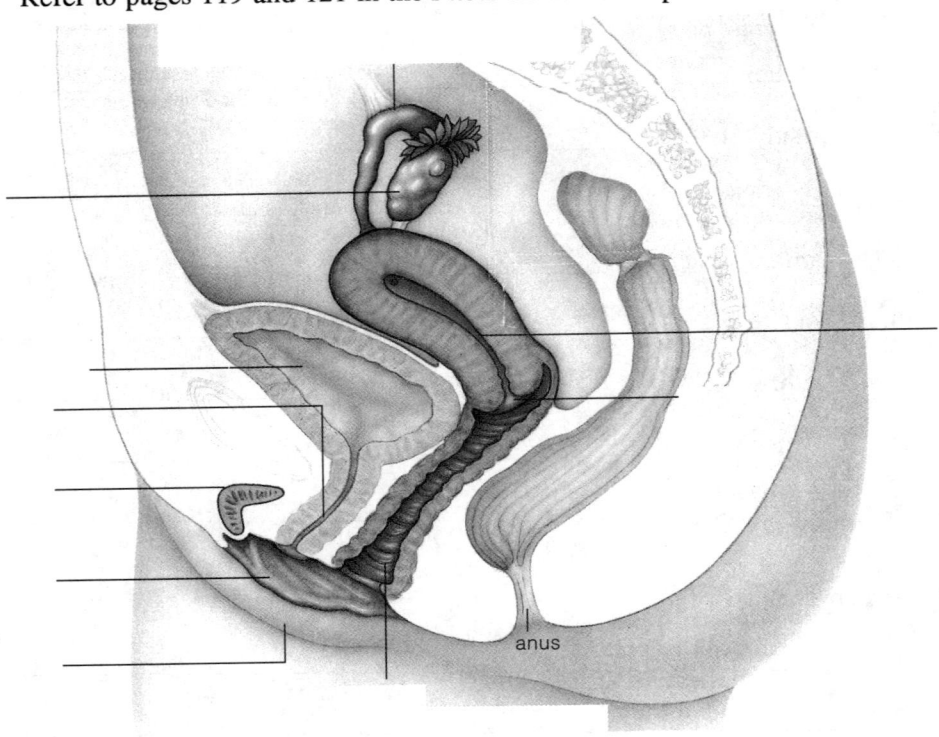

FIGURE 2

   B. Sketch the mammalian ovary, labeling a primary follicle, a secondary follicle, and the corpus luteum. (page 143, figs. 143b–f and page 144, fig. 144a)

Reproduction
Report Sheet 6

Student ID # _____

C. Trace the path of a fertilized egg from its origin in the ovary to outside the body.

1. __follicle__ 5. _____

2. __ovary__ 6. _____

3. _____ 7. _____

4. _____

D. Briefly discuss the role of the following hormones in the menstrual cycle.

1. FSH _____

2. LH _____

3. Estrogen _____

4. Progesterone _____

E. Give a function of the following structures.

   Structure   Function

1. Bartholin's Glands   _____

2. Labia   _____

3. Cervix   _____

4. Vagina   _____

5. Uterus   _____

6. Fallopian tubes   _____

F. Where does fertilization usually occur? _____

_____

_____

29 EMBRYOLOGY

LESSON OBJECTIVES

Upon completion of this laboratory exercise the student will be able to:

1. List the developmental stages of growth.

2. List the three embryological membranes and give a function of each.

3. List the three germ layers and what each gives rise to.

4. Distinguish among the morula, blastula, gastrula, zygote, embryo, and fetus if presented with drawings or descriptions of each.

MATERIALS NEEDED

1. Laboratory Manual
2. *Photo Atlas for Biology*

PREPARATION

Read the discussion which follows very carefully before attempting to complete this exercise. Also read the appropriate chapter(s) in the textbook.

DISCUSSION

Embryology is the study of the early stages of growth of an organism. This study begins with the fertilization of the egg or **ovum** by a sperm cell. Sperm cells break down the **corona radiata** surrounding the ovum or egg cell. After a single sperm cell penetrates the plasma membrane, a fertilization membrane forms enclosing the fertilized egg which is now termed a **zygote**. After fertilization, no other sperm can penetrate the egg.

Shortly after fertilization, cleavage begins in the zygote. **Cleavage** is a series of mitotic divisions in which one cell divides into 2, 2 into 4, 4 into 8, and so on. These cells are called **blastomeres**. Cleavage continues as the zygote travels from the oviduct to the uterus.

The **morula** stage is reached when the cell divisions produce a cluster of cells in a solid ball. By this time the morula has reached the uterus. The total cluster of cells is no larger than the original fertilized egg. This solid sphere of cells begins to hollow out due to the pressure of fluid accumulating between the cells. This process forms what will become the central cavity of the **blastocyst**, the next stage in development. The blastocyst contains a central cavity called the **blastocoele**. The outer layer of cells, the **trophoblast**, begins to thicken and numerous fringelike structures start to grow out into the surrounding endometrium. These outgrowths, termed **villi**, contain tiny cavities which fill with the mother's blood. Later, the embryo's blood vessels will develop within the trophoblast and will connect to the embryo via the umbilical cord. The trophoblast eventually forms the chorion and placenta which are protective and nutritive membranes. A small layer of cells, the **embryoblast**, becomes the embryo. The embryo implants itself in the uterine lining about the seventh day following fertilization where all further development takes place.

The embryo is surrounded by several fetal membranes. The outermost membrane, the **chorion**, along with the uterine endometrium makes up the placenta. The **placenta** is the organ of exchange between the mother's blood and that of the embryo, providing nutrients and oxygen to the embryo and removing wastes from the embryo. An umbilical cord develops connecting the embryo and chorion. Its function is the passage of nutrients, gases, and wastes between embryo and mother. The **amnion** surrounds the embryo. It forms a cavity filled with amniotic fluid which bathes the embryo, cushioning it against injury. The **allantois** is a vestigal membrane in the human.

The next stage in development is **gastrulation** which occurs about midway through the second week. This process is involved in the development of a gastrula from the blastocyst. Some of the cells of the blastocyst begin to invaginate to form a new cavity, the **archenteron**, which will later become the digestive tract of the embryo. As the archenteron increases in size, the blastocoel disappears. With gastrulation it is possible to identify the three primary germ layers of the developing embryo.

The external layer of cells makes up the ectoderm. The **ectoderm** gives rise to the nervous system, sense organs, skin and its outgrowths. Cells lining the archenteron make up the endoderm. The **endoderm** gives rise to the digestive tract and its outgrowths. The middle layer of cells is the **mesoderm**, which gives rise to the circulatory system, excretory system, most of the reproductive system, skeleton, and muscles.

The embryo takes shape during the third week. The primitive streak appears as a groove or line along the embryonic disc. The neural tube which later becomes the brain and spinal cord forms along the primitive streak. The brain and spinal cord are among the first organs to develop in the early embryo. The heart is formed late in the third week and is the first organ known to function. The heart begins to beat almost as soon as it is formed. Mesodermal cells form block-like structures termed somites which form bone and muscle.

Internal organs are formed during the fourth week including the liver, kidneys, lungs, and major blood vessels. Cartilage appears where the vertebral column will be, and muscles begin to form in the trunk.

The facial and neck region undergoes a series of changes resulting in slits and bulges which are called the pharyngeal gill clefts and arches. These will form the ears, jaws, mouth and part of the larynx. Optic vesicles appear where the eyes will form. Thirty somites are visible and the embryo has a tail at this stage of development. See Figure 1.

Development continues and in the fifth week, arm and leg buds appear. Facial features continue to change as nasal pits form on either side of the face. See Figure 2. The vertebral column is still composed of cartilage.

Rapid and dramatic changes occur between the sixth and eight weeks. The head and facial features become refined. The nasal pits migrate from the sides of the face to meet and form the nose. (See Figure 3 for facial features at seven weeks.) The hyomandibular clefts form the auditory canals. The eyes are formed but have no eyelids. The limb buds differentiate into toes and fingers. Reproductive organs develop early in the second month, and the sex of the embryo can be determined by the end of the eighth week; however, all of the sex organs are

WEEK 4

- yolk sac
- connecting stalk
- embryo

- forebrain
- future lens
- pharyngeal arches
- developing heart
- upper limb bud
- somites
- neural tube forming
- lower limb bud
- tail

FIGURE 1   actual length

WEEKS 5–6

- head growth exceeds growth of other regions
- retinal pigment
- future external ear
- upper limb differentiation (hand plates develop, then digital rays of future fingers; wrist, elbow start forming)
- umbilical cord formation between weeks 4 and 8 (amnion expands, forms tube that encloses the connecting stalk and a duct for blood vessels)
- foot plate

actual length

FIGURE 2

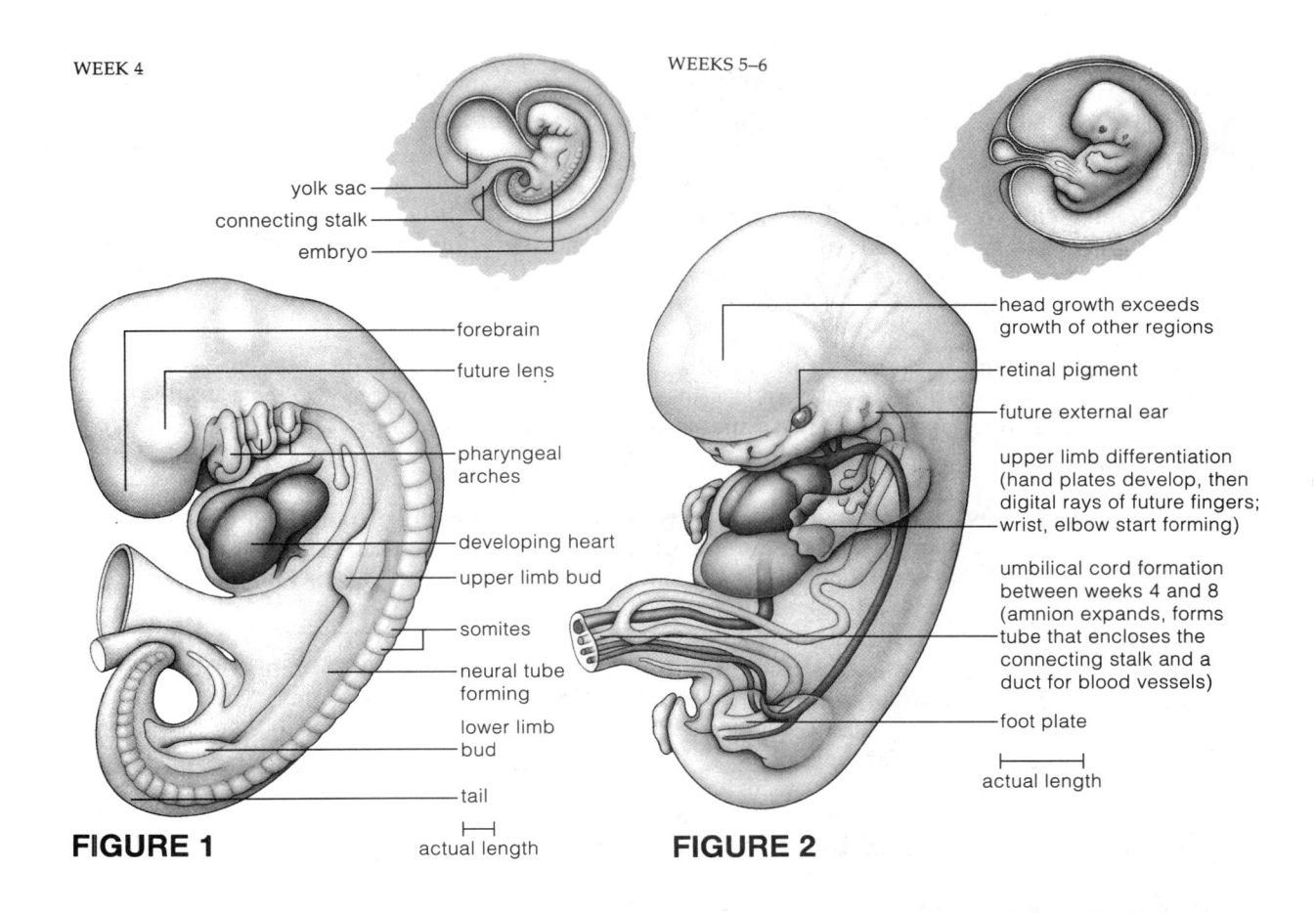

WEEK 8

final week of embryonic period; embryo looks distinctly human compared to other vertebrate embryos

upper and lower limbs well formed; fingers and then toes have separated

primordial tissues of all internal, external structures now developed

tail has become stubby

actual length

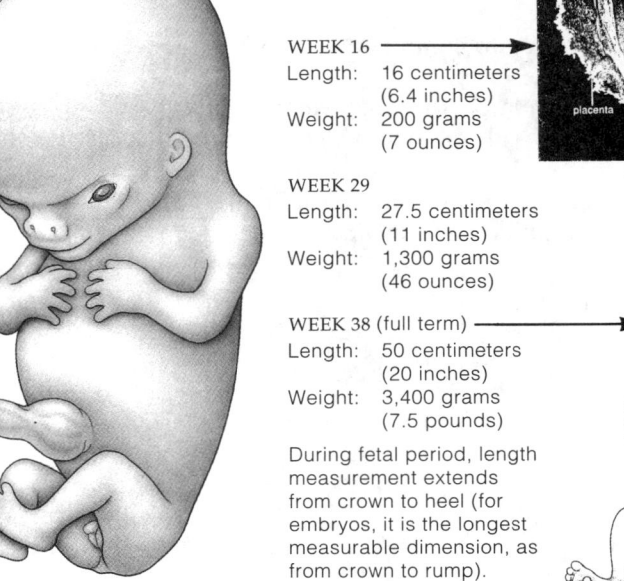

FIGURE 3

WEEK 16
Length: 16 centimeters (6.4 inches)
Weight: 200 grams (7 ounces)

WEEK 29
Length: 27.5 centimeters (11 inches)
Weight: 1,300 grams (46 ounces)

WEEK 38 (full term)
Length: 50 centimeters (20 inches)
Weight: 3,400 grams (7.5 pounds)

During fetal period, length measurement extends from crown to heel (for embryos, it is the longest measurable dimension, as from crown to rump).

FIGURE 4

Embryology   **317**

not completely developed. By the end of the eighth week all organs have been formed. Cartilage is found where the bones of the skeleton will form. The embryo is approximately one inch long but weighs less than one ounce.

The human embryo is called a **fetus** after the second month. Teeth begin development, bone is being deposited in the cartilage skeleton, and the internal organs are functioning (except the lungs and reproductive organs). Sometimes a heart beat can be detected with a stethoscope during the third month.

Refinement and growth characterize months three through nine. The fetus increases in length and weight so that by the end of the sixth month it is 14 inches long and weighs approximately 2 pounds. Eyebrow, eyelashes, lips, nails, foot- and fingerprints appear. The oil glands secrete a waxy protective covering and fine hair covers parts of the face and body.

Approximately fifty percent of birth weight is gained during the last two months. The fetus has a chance for survival when born at the end of the sixth month if proper and adequate medical facilities and care are available. The average infant weighs between 6 and 8 pounds and is 18 to 21 inches in length at birth.

Figure 4 shows the relationship of the fetus and the extraembryonic membranes at nine weeks. The developing fetus will float in the amnionic fluid, growing and moving, protected by the fluid and placenta from the outside environment until birth. With birth the embryological development of the organism is complete.

LABORATORY EXERCISE PROCEDURE

Complete the laboratory exercises and answer the questions on the lab report. When complete, submit the lab report to the instructor for grading.

LABORATORY EXERCISES

I. Use the *Photo Atlas* to complete Section I of the Report Sheet.

II. A list of terms related to embryology is provided. Arrange the terms in the correct sequence starting with the one which occurs earliest in development.

III. Locate Section III on the report sheet and complete the matching exercise using the Discussion as a guide or reference.

EMBRYOLOGY Report Sheet 1

Name _____

Student ID # _____

Campus _____

Date _____

I. Study the photos of sea star development. The first stages are similar to human development.

  A. Sketch the zygote. Label the two nuclei.

  When these two nuclei unite, the nucleus of the zygote will contain _____ (how many) chromosomes. Does either parent (mother or father) contribute more to the heredity of the offspring than the other parent? _____ (yes or no)

  B. Sketch the two-celled embryo. Label each cell.

  Do the cells grow after they divide? _____

Embryology 319

Embryology
Report Sheet 2

Student ID # _____

C. View fig. 140d, pg. 140, *Photo Atlas for Biology*

1. This figure shows a ball of cells termed a _____ which contains eight cells. Starting with a one–celled fertilized egg, how many divisions were necessary to produce this stage? _____

D. View fig. 140e. This figure shows a _____. It differs from a morula in that it is a _____ of cells.

E. View pp. 141–142 showing chicken development. Compare the embryo at 21 hours to the 72-hour embryo. Briefly discuss the major changes in development.

_____

_____

II. The following list consists of terms related to embryology. Arrange the stages and/or events in the correct sequence by placing the letters in the appropriate numerical order starting with the one which occurs earliest in development.

a. morula
b. organ formation
c. zygote
d. optic vesicles
e. infant
f. heart forms
g. fetus
h. fingers and toes
i. cleavage
j. villi appear

_____ 1.   _____ 6.

_____ 2.   _____ 7.

_____ 3.   _____ 8.

_____ 4.   _____ 9.

_____ 5.   _____ 10.

320   Exercise 29

Embryology
Report Sheet 3

Student ID # _____

III. Matching: Match the structure in Column A to the correct description in Column B. Place the appropriate letters in the blanks preceding each structure.

Column A

_____ 1. Nasal Pit
_____ 2. Cartilage skeleton
_____ 3. Somite
_____ 4. Trophoblast
_____ 5. Ectoderm
_____ 6. Placenta
_____ 7. Zygote
_____ 8. Cleavage
_____ 9. Endoderm
_____ 10. Neural tube
_____ 11. Mesodermal cells
_____ 12. Optic vesicles
_____ 13. Morula
_____ 14. Fetus
_____ 15. Blastocyst

Column B

A. fertilized egg
B. series of mitotic divisions
C. hollow ball of cells
D. solid ball of cells
E. becomes nostril
F. embryo after two months
G. forms chorion and placenta
H. forms brain and spinal cord
I. forms somites
J. bone and muscle form from these
K. bony skeleton
L. eyes
M. nervous system, sense organs, skin
N. digestive tract
O. protects developing fetus

30 ECOLOGY: POPULATIONS AND COMMUNITIES

LESSON OBJECTIVES

Upon completion of this laboratory exercise the student will be able to:

1. Define the terms population, community and ecosystem.

2. Use a growth curve to describe the changes in a population and explain the cause for each stage of growth.

3. Describe the human growth curve, noting periods of growth and stability, and give reasons for those changes.

4. Determine the density of a population and suggest reasons for changes in density of a population.

MATERIALS NEEDED

I. Population Growth
    a. seeds (bean or pea)
    b. paper cups
    c. container to hold the seeds

II. Population Density
    a. meter stick or yard stick
    b. cord or string
    c. 4 large nails

PREPARATION

Read the discussion which follows carefully before attempting to complete the exercise. Also read the appropriate chapter(s) in your textbook.

DISCUSSION

Ecology is the study of the relationship between living organisms and their environment. An **ecosystem** is all of the living community (community of plants and animals) and its **abiotic** (nonliving) environment. Abiotic factors which affect the community include temperature, type of soil, amount of sunlight, and amount of moisture. Life is most abundant with high temperatures, plenty of rain, sunshine all year, and fertile soil.

A **community** is all of the plant and animal populations living within a geographical area. A **population** is all of the members of a given species within a given area. Each population of organisms occupies a certain place within the community, called its **habitat**. Each species also has its own ecological **niche** or "job" within the community.

A population within the community may be stable in that its numbers remain about the same year after year. If the population undergoes large yearly changes in numbers, then it is said to be unstable. Many factors contribute to the stability of a population. The available food supply, the birthrate of the species, and the number of predators of the species influence the stability of the population. The addition of new individuals to the population (immigration) or the loss of individuals from the population by death or immigration affect the population's stability.

Biologists measure the size of a population by determining its density. **Density** is determined by counting the number of organisms of one species in a given area or volume of space. The total number of organisms is then divided by the total space. Density is expressed as the number of organisms in the unit area or volume.

$$\text{Density} = \frac{\text{Total number of organisms}}{\text{total area}}$$

By determining the density of a population at various times, the biologist can determine whether the population is increasing, decreasing or remaining the same. That information can indicate how the area is changing in terms of such things as food supplies for the population or the carrying capacity (number of organisms an area can support) or the area. Because plants are nonmotile, it is much easier to determine the density of a plant population.

Populations change in number over a period of time. The communities in which they are found also change. This is called **ecological succession**. The final community in a succession is the **climax community**.

LABORATORY EXERCISE PROCEDURE

Carefully follow the directions for each of the following exercises, recording observations, conclusions, and answers on the appropriate sections of the report sheets. When the exercise is complete, submit the completed report sheets to the instructor for grading.

LABORATORY EXERCISES

I. Population Growth

    A. Objective
        1. Demonstrate the rate of population growth
        2. Project the increase in population growth in 50-year intervals

    B. Materials
        1. A container large enough to hold a large number of seeds
        2. Bean or pea seeds
        3. Nine paper cups

    C. Procedure
        1. Place four seeds inside a large container. The seeds represents the earth's population in the year 1575. Each of the four seeds represents 100 million people or a total of 400 million people in 1575.
        2. On the paper cups write the years 1625 through 2025 in 50 year increments. Place one seed in cups one (1625), two (1675), three (1725), and four (1775), respectively. Cup five (1825) should hold two seeds; cup six (1875) three seeds; cup seven (1925) seven seeds; cup eight (1975) twenty seeds; and cup nine (2025) 43 seeds.

3. Add the contents of the second cup (1625) to the initial container, recording the population increase and total population in the chart in the lab report.
4. Repeat step three for each cup, recording population increase and total population.
5. Graph the results in the lab report.

II. Population Density

    A. Materials
1. Meter stick or yard stick
2. Cord or string
3. 4 large nails

    B. Procedure
1. On a lawn, pasture or park, randomly select a one meter or one yard square area. Use the four nails for the corners of your square. Tie the cord to each nail until you have a square outlined.
2. Staring along one edge of the plot, place the meter stick or yard stick across the plot and count the dandelion and henbit plants that touch the meter or yard stick. If your area does not have these plants, substitute others such as clover or wild rye grass. Be sure that the plants you use are easily visible.
3. Continue to move the meter or yard stick across the plot, moving toward the opposite edge of the plot. Count the plants touching the stick as you move it across the plot. Count each plant only once.
4. Record the number of dandelion and henbit plants in the table in your lab report.
5. Repeat the procedure for three more plots, recording the number of plants.
6. Go to a different area and repeat the experiment using four trials. If you used a lawn, try a pasture or park. Record your results in the table in the lab report.
7. Determine the average number of dandelion and henbit plants for each set of data.
8. Answer the questions in the lab report.

ECOLOGY: POPULATION AND COMMUNITIES

Report Sheet 1

Name _____

Student ID # _____

Campus _____

Date _____

I. Population Growth

Year	Increase in Population	Total World Population (in millions)
1575		
1625		
1675		
1725		
1775		
1825		
1875		
1925		
1975		
2025		

Total World Population in millions

(Graph your results)

8000
7000
6000
5000
4000
3000
2000
1000

1575 1625 1675 1725 1775 1825 1875 1925 1975 2025

326   Exercise 30

Ecology: Population and Communities Student ID # _____
Report Sheet 2

A. What trend in population growth can be determined from the graph? _____

_____
_____
_____

B. How often does the world's population double in size? _____

_____
_____
_____

C. Based on the graphed information, project the world population size in the year 2000.

_____

D. Predict some of the effects population growth will have on the natural resources of the earth.

_____
_____
_____
_____

II. Population Density

Location 1

Plot	Plant 1	Plant 2
1		
2		
3		
4		
Total		
Average		

Ecology: Population and Communities Student ID # _____
Report Sheet 3

Location 2

Plot	Plant 1	Plant 2
1		
2		
3		
4		
Total		
Average		

A. Which location had the higher population density of dandelions (plant 1)?

_____

B. What physical and biological factors control the population density of each type of plant in location 1?

_____

_____

_____

C. If you were to return to the exact same plot one week later, would the two plants be the same, show an increase, or show a decrease?

_____

Explain. _____

_____

_____

D. Would the population densities of either plant change after a month? _____

After another season? _____ Explain. _____

_____

_____

Ecology: Population and Communities   Student ID # _____
Report Sheet 4

E. What methods would you have to use to estimate animal populations?

_____

_____

_____

Why?_____

_____

_____

31 ECOLOGY: ECOSYSTEMS AND THE BIOSPHERE

LESSON OBJECTIVES

Upon completion of this laboratory exercise the student will be able to:

1. Describe the general pattern of energy flow through an ecosystem.

2. Construct a food web, using the terms producer, consumer, and decomposer.

3. List the six major biomes and discuss the abiotic factors associated with each biome.

4. List representative vegetation in each biome.

5. Determine the adaptations of plant life to the abiotic factors of each biome.

MATERIALS NEEDED

1. Laboratory Manual
2. Textbook
3. Pen or pencil

PREPARATION

Read the discussion which follows carefully before attempting to complete the exercise. Also read the appropriate chapter(s) in your textbook.

DISCUSSION

An **ecosystem** is all of the living community (community of plants and animals) and its **abiotic** (nonliving) environment. Abiotic factors which affect the community include temperature, type of soil, amount of sunlight, and amount of moisture. Life is most abundant with high temperatures, plenty of rain, sunshine all year, and fertile soil.

A **community** is all of the plant and animal population living within a geographical area. Within the community, populations (all members of a given species within a given area) feed upon other populations, forming a food web. A **food web** is a complex feeding system containing several food chains. The first link in any food chain is the producer. A **producer** is an autotroph. An **autotroph** is an organism that manufactures its own food. Producers are always green plants. Succeeding links on the food chain are called **consumers** because they eat the producers or other consumers below them on the food chain. Consumers are **heterotrophs** because they must get their food from an outside source; they cannot photosynthesize. The last link on the food chain are the **decomposers**. Decomposers live off of dead plants and animals, causing them to decay.

Food chains can be represented as a pyramid. Each level is smaller than the level below it. Food energy is always less on a level than the level below it because as energy is transferred, most of it is lost to the environment as heat. The following figure illustrates a food chain pyramid.

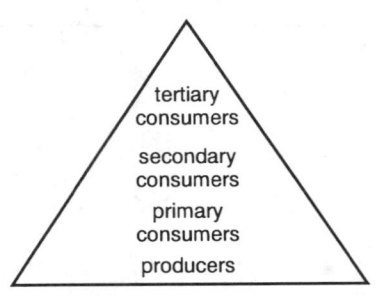

```
       /\
      /  \
     /tertiary\
    /consumers \
   /            \
  / secondary    \
 /  consumers     \
/   primary        \
/   consumers       \
/    producers       \
----------------------
```

A primary consumer usually eats only producers. An example is a herbivore like the cow. Secondary consumers can be either plant–eaters (herbivores), flesh–eaters (carnivores), or both (omnivores). Tertiary consumers would fall into the same categories as secondary consumers. In some cases, primary consumers may also be carnivores.

The plant and animal species within a community are dependent upon the type of climate in which they reside. Each geographical region with its own temperature range and average rainfall is associated with a particular type of vegetation. Each of these regions is called a **biome**. The boundaries of biomes are not sharply defined but may be divided into tropical rain forest, temperate forest, desert, grassland, tropical savannah, tundra, and ocean.

Climax communities within a biome remain relatively stable. The carrying capacity of the biome permits only a certain number of organisms to grow. Whenever the ecosystem is altered in some way, change occurs in the community. Overpopulation is a major problem. The human population is facing a severe food shortage as well as an energy shortage due to overpopulation.

LABORATORY EXERCISE PROCEDURE

Carefully read all of the discussion material before attempting the following exercises. Carefully complete each exercise recording the required information on the report sheet. When complete, submit the reports to the instructor for grading.

LABORATORY EXERCISES

I. Food Web

   A. Using a coin, draw ten circles on your lab report. Space the circles randomly.
   B. In each circle print the name of an organism from the following chart. Use each name only once and arrange animals with similar eating habits so that they are not clustered together. For example:

Producer:	tree, clover
Primary Consumer:	bees, squirrels, beetles, mice
Secondary Consumer:	woodpeckers, cats, owls
Scavenger:	termites

- C. Draw an arrow from each organism to every other organism that may depend upon it as a source of food. Some organisms may have several arrows pointing to and away from themselves.
- D. Assume that putting an X in a circle means that the members of that species have been removed from the community. Put an X in the circle labeled beetles. Then draw small x's over each arrow pointing *away* from the beetle circle.
- E. Answer the questions in the lab report.

II. Biome Lab

- A. Go to a nursery or botanical garden to observe the following species of plants. If the plant is not available, ask an attendant about a substitute and observe it.

    Reindeer moss
    branch of a conifer such as fir, pine, spruce
    branch of a deciduous tree such as American beech, sugar maple, oak,
        willow, hickory
    bluestem and buffalo grass
    rubber plant
    bromeliad
    philodendron
    potted cactus
    resurrection plant

- B. Answer the questions in your lab report.

ECOLOGY: ECOSYSTEMS AND THE BIOSPHERE Report Sheet 1

Name _____

Student ID # _____

Campus _____

Date _____

I. Food Web

   A. Model of a Food Web (Place your circles and connecting lines here. Be sure to label your circles.)

   B. Answer the following questions:

      1. List the animals that lost a food source when the beetles were removed from the community. Next to each of the animals list the sources of food still available to it.

         _____

         _____

         _____

         _____

Ecology: Ecosystems and the Biosphere   Student ID # _____
Report Sheet 2

3. For each animal that lost a food source, list the animals that depend upon it for food.

_____
_____
_____
_____

II. Biomes

　A. Answer the following questions:

　　1. Which of the following abiotic factors is primarily responsible for the formation of a desert? soil quality, light, precipitation, temperature, wind (Put the correct answer in the space below.)

_____

　　2. In which biomes would you expect to find the greatest varieties of plants and animals?

_____

Explain your choices. _____
_____
_____
_____

　　3. How is the rootlike system of the reindeer moss well adapted for the tundra?

_____
_____

　　4. How does the shape of a conifer tree and the shape of its leaves adapt it to the taiga biome?

_____
_____
_____

Ecology: Ecosystems and the Biosphere   Student ID # _____
Report Sheet 3

5. How is the conifer leaf adapted to a dry biome? _____
_____
_____

6. Which type of grass, bluestem or buffalo, would you expect to find in an area of greater precipitation?
_____

7. How are the broad leaves of the rubber plant adapted for the tropical rain forest?
_____
_____
_____

8. How does a bromeliad obtain water in a tropical rain forest? _____
_____
_____

9. Why does a cactus have such a thick outer covering and such a thick stem?
_____
_____

In which biome would you expect to find the cactus? _____

10. Which would be able to absorb more sunlight? a conifer or a deciduous leaf?
_____ Explain your answer. _____
_____
_____
_____

336   Exercise 31

Ecology: Ecosystems and the Biosphere   Student ID # _____
Report Sheet 4

III. Match the item in Column A with its description in Column B. Answers will be letters.

COLUMN A

_____ 1. ecology

_____ 2. abiotic

_____ 3. community

_____ 4. population

_____ 5. food web

_____ 6. producer

_____ 7. autotroph

_____ 8. consumer

_____ 9. decomposer

_____ 10. herbivore

COLUMN B

a. producer

b. plant–eater

c. contains several food chains

d. the study of the relationship between living organisms and their environment

e. live off dead plants and animals

f. first link in a food chain

g. all members of a given species within a given area

h. nonliving

i. eat producers

j. all plants and animals living in a geographical area

GENERAL BIOME INFORMATION

Biome	Annual Rainfall	Average Temp.	Growing Season	Soil Type	Other Information
Tundra	Very dry, less than 12 inches annually	$-30°$ to $10°$ C	very short, less than 60 days	few minerals; soil frozen except for uppermost yard	long, cold winters; short summers
Taiga	dry, 20 to 40 inches annually	$-30°$ to $20°$ C	short, 60 to 150 days	poor, acidic soil; lacking nutrients	long, cold, snowy winters; warmer summers than tundra
Temperate Forest	moderate rainfall; 30 to 50 inches annually	north: $-12°$ to $27°$ C; south: $15°$ to $30°$ C	moderate, 180 days or more	good soil, with many nutrients	long, warm summers; north cold winters; south, warmer winters
Grassland	dry, 10 to 30 in with irregular precipitation patterns and drought	wide temp. ranges between temperate forest and desert	moderate, 180 days or more	very rich soil; easy to grow crops except during drought	long, warm summers; winters severe in some areas
Tropical Rain Forest	very wet, over 79 inches annually	constant, high temps. $24°$ to $26°$ C	growing season year round	poor soil; minerals, nutrients washed away by rain	warm and humid all year
Desert	very dry, less than 10 in with irregular rain patterns and drought	cool desert, $10°$ C; hot desert, $20°$ C	depends on type of desert	poor, sandy soil	depends on type and location of desert

32 ECOLOGY: THE HUMAN FACTOR

LESSON OBJECTIVES

Upon completion of this laboratory exercise the student will be able to:

1. Identify major air pollutants and their sources.

2. Discuss the effects of these pollutants on human and plant health.

3. Identify sources of water pollution and describe the effects of water pollution on living organisms.

MATERIALS NEEDED

I. Air Pollution

    1. Color Plates of Smoker's Lung and Emphysema Lung*

II. Water Pollution

    1. Masking tape for labels
    2. Small plastic bags (small sandwich bags are best)
    3. Paper towels
    4. Dried seeds (pinto bean, lima bean, etc.)
    5. Small containers in which to soak the seeds
    6. Water
    7. Liquid detergent
    8. Powdered detergent dissolved in water
    9. A biodegradable detergent dissolved in water (read detergent label to determine biodegradability)

    *This item is found in your lab kit.

PREPARATION

Read the discussion which follows carefully and the appropriate chapter(s) in your textbook.

DISCUSSION

Man has introduced many harmful substances into his environment. The ultimate effects of many of these substances may not be known for years. These include pesticides, wastes from atomic reactors, mountains of nonbiodegradable garbage, asbestos, and excessive amounts of noxious fumes. In addition, introduction of new species into an environment where they compete with existing species, often results in the complete disruption of a balanced environment.

In addition to polluting our environment, we are rapidly depleting our nonrenewable resources, such as petroleum, and overusing the renewable ones.

Air pollution and water pollution are major concerns for the environment. Pollutants such as gases or airborne particles like volcanic dust all contribute to the contamination of the air. Water pollutants include biological agents, chemicals and heat. Sewage is a major water pollutant, as is heat from factories. Water sources are being "poisoned" to the extent that the water is no longer usable. The smog from airborne pollutants has caused many health problems in the elderly, young children and people with respiratory problems.

For the most part the clean-up drives and anti-litter campaigns have dealt primarily with aesthetic issues. The informed citizen recognizes that the issues to be faced now, not sometime in the future, are much more serious than merely picking up litter on the roadside. Poisons which destroy entire species must be identified and, if possible, eliminated because an extinct animal or plant species is just that—gone forever. When we eliminate one species we may increase the numbers of individuals in another. If we introduce a new species into a given environment we may cause untold ecological and economical damage. The case of Thomas Austin's rabbits is a classic example. Austin imported twelve pairs of European rabbits to Australia in 1859 and released them on his ranch. Six years later he is reported to have killed twenty thousand rabbits! Australia had an estimated one billion rabbits by 1953 despite all attempts at control. Economically this meant fewer and fewer sheep could be raised as five rabbits eat as much grass as one sheep.

LABORATORY EXERCISE PROCEDURE

Perform the indicated exercises and prepare a written report based on one of the listed topics. When complete, submit the report sheets and the ecological issues report to the instructor for grading.

LABORATORY EXERCISES

I. Air Pollution

   A. Look at the Color Plates in the Lab Kit
   B. Answer the questions in the lab report

II. Water Pollution

   A. Objective—Compare germination times of seeds soaked in various solutions to determine the effects of pollutants on germination

   B. Procedure
      1. Label the containers water, liquid detergent, powdered detergent, and biodegradable detergent.
      2. Place 10 seeds into each jar and cover completely with one of the solutions.
      3. Soak the seeds for 24 hours.
      4. After the 24-hour period, remove the seeds from each container (keep them separate) and place them between two layers of paper towels. Moisten the paper towels and slide them into the plastic bags. Be sure to label the plastic bags.
      5. Examine the seeds the next day for signs of germination. The appearance of a small root or sprout indicates a germinating seed.

6. Record the number of seeds germinating in each group and record that information in the chart in the lab report. Then determine the percentage of germinating seeds in each group.

III. Ecological Issues Report

A list of activities is provided below. Select one and write a report on it. The laboratory instructor may have other activities to suggest. Consult the laboratory instructor and decide on a project. The instructor may give hints and advice as to how to proceed and/or where materials can be obtained This laboratory exercise can be as rewarding and enlightening as the effort extended. The report should be a minimum of three pages in length with appropriate bibliography. Illustrations such as photographs, graphs, charts, and/or maps should be included as needed. The paper *must be typewritten*. Handwritten papers *will not be accepted* and papers with less than three pages will not receive any credit.

ACTIVITIES (Select One)

1. Identify a local area or areas where pollution of the environment is occurring or has occurred recently. Fully describe the nature of the pollution. Describe the changes which have occurred, and explain who or what was affected and how. Determine what, if anything, has been done to correct the situation. Fully describe what has been done, if anything, and evaluate the results. Give suggestions for improving the conditions.

2. Select an environmental area to study over a specific period of time. Choose an area which is easily accessible so that you can make as many trips to it as necessary to complete your research. Suggestion: Mark off a one square meter or one square yard area using wooden stakes and twine. Include the number of different species, the number of individuals in one species, the dominant species, microhabitats, and notable physical factors present in the survey area. If you disturb a rock or log, replace it. Describe the ecology of the area in as much detail as possible using sketches or photos to illustrate the report. Carefully describe all changes observed during the period of study.

3. If you live in an area in which a natural disaster such as flooding, wind damage, tornado, hail, drought, earthquake, landslide, or range fire has occurred recently, determine the extent of the damages to the region both ecologically and economically. Try to determine which plant and animal species suffered the greatest distress. What efforts are being made to restore the region?

4. Study your own habits or those of someone you know well. Are your attitudes and habits consistent with sound ecological principles? Describe any that are particularly noteworthy. Do your actions affect anyone else? Are you conservative with resources and energy? Determine and report how much garbage you generate in a week by collecting yours in a separate container and weighing it. Determine what efforts your community is making to dispose of its wastes. Obtain permission to visit a landfill, water treatment plant or city–county health facility. Describe what you see in as much detail as possible. Find out if air in your area is monitored for pollution. Describe and give results if you can obtain them. Any one of these areas can provide ample materials for an interesting research project.

5. Determine what types of legislation exist to protect the citizens of your state against various pollutants and environmental stresses such as strip mining, air

pollution, and water pollution. Do local industries comply with these laws? What agency is charged with enforcing environmental protection laws? Extend this research to include national laws and acts aimed at protecting our environment.

6. Noise bombards us from every angle. It seems to disturb some people more than others. Report on persistent sources of unusually loud noises in your area. Do they affect many people? Are there ways to prevent these noises or at least lower the noise factor? Are there local laws to control noise? Who enforces these, if any? Is noise pollution an invasion of rights?

7. Insecticides and pesticides have been in wide use since the first application of DDT in the 1930's. The buildup of DDT in the environment and its effect on animal populations is an interesting and thought-provoking story. Research DDT from its entrance into the United States to the present.

8. Food pollution may be a new term to you. It is disturbing to most humans because our health and very survival depend on our food supply. Research food production and processing. Give attention to practices which result in chemicals being added to food supplies in the form of coloring agents, flavor enhancers, vitamins, minerals, hormones, enzymes, texturizers, antibiotics, and fillage. Some food processing methods result in losses of minerals and vitamins. Give examples of these. Do economic factors and customer demand account for the need for food additives? If so, how?

9. As man rapidly depletes his natural resource, he must learn quickly how to manage remaining natural resources. Finding alternate energy sources is a particularly pressing problem. Prepare a report on the energy crisis. Include such factors as demands caused by increasing population, depletion rate of fossil fuels, loss of open spaces, blackouts, destruction of forests, recycling, new energy sources, and environmental impact of increased use of alternate energy sources. Suggest changes in lifestyles and attitudes. Survey sources of energy and present depletion rates, distribution or resources, and political factors which contribute to the energy crisis.

10. Select an appropriate book on ecology or ecological issues to read and report on. You must clear the book content with your lab instructor before it will be accepted.

ECOLOGY: THE HUMAN FACTOR

Report Sheet 1

Name _____

Student ID # _____

Campus _____

Date _____

I. Air Pollution

A. Look at the normal human lung tissue and sketch what you see.

B. Observe the lung section showing emphysema. Describe the differences in this picture and normal lung tissue.

_____
_____
_____
_____
_____

C. Observe the smoker's lung. Describe its differences from normal lung tissue.

_____
_____
_____
_____
_____

Ecology: The Human Factor　　　　Student ID # _____
Report Sheet 2

D. Why is smoking a problem? _____
_____
_____

Do you think that secondary smoke can cause some of the same changes seen in the lung tissue? Why or why not?

_____
_____
_____
_____

II. Water Pollution

　A. The effect of a pollutant on seed germination

Solution Used	Number of Seeds	Number of Germinating Seeds	Percent Germination
Water	10		
Liquid Detergent	10		
Powdered Detergent	10		
Biodegradable Detergent	10		

　B. Answer the following questions:

　　1. What effects do liquid detergent have on seed germination? _____
_____
_____

Ecology: The Human Factor
Report Sheet 3

Student ID # _____

Powdered detergents? _____
_____
_____

Biodegradable detergents? _____
_____
_____

2. Of the three detergents tested, which one would be the better for the environment?

_____ Why? _____
_____
_____

3. Most people pollute the water system in some way. How do you contribute to water pollution?

_____
_____
_____

What could you do to reduce water pollution? _____
_____
_____
_____

4. What could you do to conserve water? _____
_____
_____
_____

33 BEHAVIOR

LESSON OBJECTIVES

Upon completion of this laboratory exercise the student will be able to:

1. Give a definition for behavior.

2. Define tropism and give an example.

3. Distinguish among taxis, reflex, and instinct.

4. List the types of learned behavior and give an example of each.

5. Answer the questions in the lab report.

MATERIALS NEEDED

1. Laboratory Manual
2. Pen or pencil
3. Bean seeds (Try to get bean seeds from a seed store or nursery; dried lima beans or kidney beans may be used, but *may not* give the desired results.)
4. Paper towels
5. Two small plastic bags (sandwich bags will do)
6. Straight pins
7. Tape or staples
8. Two pieces of cardboard at least as large as the plastic bag when the board is folded in half

PREPARATION

Read the discussion which follows carefully before attempting to complete the exercise. Read the appropriate chapter(s) in your textbook as well.

DISCUSSION

Behavior is the way in which an organism reacts in response to an internal or external stimulus. The major purpose for behavior is to return the organism to equilibrium or homeostasis. In other words, behavior is what an organism uses for survival, be it food–getting, defense, reproduction, etc.

Both plants and animals show responses that can be termed behavior. Plant growth responses to stimuli are called **tropisms**. These tropisms are innate and inflexible responses. Examples of tropisms include the following:

Phototropism	—	response to light
Geotropism	—	response to gravity
Heliotropism	—	response to the sun
Thermotropism	—	response to temperature
Thigmotropism	—	response to touch

A **nastic** response in a plant is independent of the direction of the stimulus. **Thigmonasty** is an example of a nastic response by the mimosa plant, which closes up its leaves and folds to the ground when touched or shaken. This response is caused by changes in the water volume of certain cells within the plant and is a non-directional external stimulus response.

Innate animal behavior which is inherited may be classified as a taxis, reflex, or instinct. A **taxis** is a very simple response of the entire body to a directional stimulus. Movement toward light as exhibited by some insects is an example of positive phototaxis. There are also other taxis such as chemotaxis and geotaxis. The **reflex** is a response of only a part of the body to a stimulus. A reflex response occurs in most organisms with a central nervous system. Blinking the eye to avoid a threat is an example of a reflex. **Instinct** is a complex inborn behavior pattern. Sometimes it is difficult to distinguish instinct from learned behavior.

Learned behavior is a response modified or changed as a result of experience while innate behavior is an inherited response pattern. Innate behavior can be modified to a certain extent; however, if proper response is necessary for survival with no time for learning, then the organism born with the appropriate innate behavior is the survivor.

Organisms interact with other organisms. **Social behavior** refers to the interaction of organisms with members of their own species. Social behavior may be innate, learned, or both. Courtship and territoriality are two types of social behavior. **Courtship** is the group behavioral patterns accompanying mating. Courtship serves two functions: to recognize a "mate" of the opposite sex in the same species and to correlate egg production with mating.

The establishment of a local area for functions such as mating, living, or feeding is termed **territoriality**. Many species of animals demonstrate territoriality. A very common example of the establishment of a territory occurs when you walk a male dog. The dog will smell all trees, bushes, etc., and then mark them with urine to announce his area. A territory's boundary is the line at which the male's urge to defend his area is equal to his urge to flee. The territoriality behavior response lessens overcrowding in a habitat, assures an adequate food supply, and tends to keep the population in check.

Within a group of animals, social order or ranking system insures the maintenance of order within the group. A herd of horses will show ranking or a "pecking order." There is a chain of command or **dominance hierarchy** within the group. This order determines who will eat first, drink first, etc. The order is not changed unless a new horse is added to the group. Then a new order of social dominance will be formed with the new horse fitting in at his level of dominance. In an ordered group, the top horse will pick on number two, number two will pick on number three, and so on until the lowest horse in the social or pecking order is attacked. The lowest animal will not fight back.

Learning accounts for a large amount of behavior in animals. **Habituation** is a type of learned response in which the organism learns to ignore constant or irrelevant stimuli. For example, you have learned to ignore television noise when you are trying to study. **Imprinting** is learned behavior in which a newborn organism such as a gosling will attach itself to whatever it sees move as if the object were its mother. **Conditioning** is another type of learned behavior which depends on reinforcement by punishment or reward. A dog will salivate when he smells food.

If a bell is rung at the same time the food is placed before the dog, after a time the dog will produce saliva upon hearing the bell because he is conditioned to expect food.

A more advanced type of learning is **trial and error**. The highest form of learning is **insight** or **reasoning** in which previously acquired knowledge is used to solve new problems. This type of behavior is generally associated only with humans and occasionally a few higher primates.

Humans exhibit both innate and learned behavior patterns. One particular behavior pattern shared by all animals is personal space. **Personal space** is a measure of the closeness an animal is willing to tolerate between itself and others of its species in a given situation. When personal space is violated, the animal will either fight or flee, depending upon the circumstances. Usually adjustments are made to reestablish personal space when two or more animals invade each other's space.

LABORATORY EXERCISE PROCEDURE

Carefully read the directions for each laboratory exercise. Complete each exercise recording the required information on the report sheets. When complete, submit the report sheets to the instructor for grading.

LABORATORY EXERCISES

I. Plant Tropisms

    A. Objective—to observe the plant response to gravity or geotropism

    B. Materials—see the first page of this exercise

    C. Procedure

        1. Soak eight bean seeds in water overnight. Suggestion: Start several groups of seeds in case some do not germinate.
        2. Fold one piece of cardboard in half so that there are two equal sides the size of the plastic bag. The cardboard should be stiff enough to stand by itself. Label this cardboard as experimental group I.
        3. Tape or staple a folded paper towel to one side of the folded cardboard.
        4. With straight pins, attach four seeds (equal distances apart from each other) to the paper towel and cardboard about half way from the fold in the cardboard. Be sure to place the seeds as indicated in the diagram.

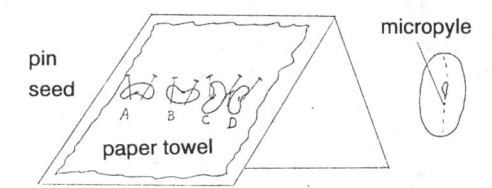

FIGURE 1

In positioning seeds C & D, look for a small hole (called a micropyle) which appears on the side of each bean seed. The micropyle should be toward the top for seed C; toward the bottom for seed D.

        5. Now place some water in the bottom of one of the plastic bags and place the cardboard containing the seeds in it. Make sure the paper towel touches the water but *do not* cover the seeds with water.

6. Repeat steps 2–5 with the second piece of cardboard and four more seeds. Label this cardboard as experimental group II. *Suggestion: Start several group II set–ups in case some do not germinate.*
7. Set both experimental groups in a place where they won't be disturbed and observe any growth. Be sure to add water to the bags when needed. *Do not let the paper towels go completely dry.* (Each group should receive the same amount of light.)
8. After root growth has been established, turn cardboard II containing seeds upside down. (Cut off the folded back of the cardboard and just prop the cardboard up.) Replace the plastic bag as in step 5. Observe any changes in root growth direction and record the observations on the report sheet.

II. Habits

　A. Objective—to observe some common habits and how they are learned

　B. Materials

　　1. A shoe with laces
　　2. A buttoned shirt
　　3. A piece of bubble gum

　C. Procedure

　　1. Choose one of these habits to discuss:
　　　a. Tying a shoelace
　　　b. Buttoning a shirt
　　　c. Blowing a bubble with bubble gum
　　　d. Whistling
　　2. Write out a step–by–step procedure for doing one of the above. Be sure to include each *exact* step.
　　3. Give the procedure to another person and have that person attempt the habit using *only* the written procedure and assuming nothing.
　　4. Answer the questions in the lab report attaching the instruction or procedure sheet (from II C 2) to the report sheet.

III. Spacing and Behavior

　A. Objective—to conduct some experiments that relate to personal space in humans

　B. Procedure

　　1. Without speaking to the person, sit very close to someone you do not know. Record your reactions and the person's reactions in the lab report. (Example: Sit at a table in the library with a stranger.)
　　2. Now sit close to a friend and record your reactions and your friend's reactions.
　　3. Repeat the same experiment, but substitute a family member such as husband, wife, children, parents, etc. Record your reactions and their reactions.
　　4. OPTIONAL EXERCISE
　　　Go to a public place such as a library or a restaurant and sit (a) directly opposite, (b) adjacent to, (c) diagonal to different people. Record their reactions and your reaction.

BEHAVIOR

Report Sheet 1

Name _____

Student ID # _____

Campus _____

Date _____

I. Plant Tropisms

    A. Compare the growth of the seeds on the cardboards I and II. (Include observations relating to the <u>direction</u> and <u>amount</u> of growth.)

        Experimental Group I   Experimental Group II

        _____   _____

        _____   _____

        _____   _____

        _____   _____

    B. Define:

        1. Behavior _____

        _____

        2. Tropism _____

        _____

        3. Reflex _____

        _____

        4. Taxis _____

        _____

        5. Instinct _____

        _____

Behhavior
Report Sheet 2

Student ID # _____

C. Use the following terms to fill in the blanks:

geotropism   thigmotropism   tropism
learned behavior phototropism

1. A _____ is a plant growth response to a directional stimulus.

2. Response of a plant to light is termed _____.

3. The effect of gravity on plant growth is called _____.

4. A plant response to touch is _____.

5. A response which is modified or changed as a result of experience is a

   _____.

Five types of learned behavior are discussed in this lab. Match each type listed with its example by placing the correct letter in the blanks preceding the numbers.

Learned Behavior		Example
_____ 6. Habituation	a.	using previous knowledge to solve new problems
_____ 7. Imprinting	b.	ignore TV or radio noise when studying
_____ 8. Conditioning	c.	newborn attaching itself to a moving object
_____ 9. Trial and Error	d.	dog salivates when bell is rung
_____ 10. Insight	e.	learning to tie shoes

11. _____ is a plant growth response to temperature change.

12. A non–directional external stimulus response is a _____.

13. A non–directional external response to shock is called _____.

14. _____ is the establishment of an area for living or feeding.

15. The measure of the closeness an animal is willing to tolerate between itself and others of its species is termed

    _____.

352 Exercise 33

Behhavior
Report Sheet 3

Student ID # _____

II. Habits

1. Was there any difficulty listing the steps in the habit selected? _____

    Why or why not? _____

2. Did the partner successfully complete the habit selected? _____

    Why or why not? _____

3. What is the best way to learn a habit? _____

    _____

III. Spacing and Behavior

1. What was the response from each individual when:

    a. Sitting next to a stranger? _____

    b. Sitting next to a friend? _____

    c. Sitting next to a family member? _____

2. How did you feel when you:

    a. Sat next to a stranger? _____

    b. Sat next to a friend? _____

    c. Sat next to a family member? _____

3. OPTIONAL EXERCISE

    If you went to a public place and seated yourself (a) directly opposite, (b) adjacent to, and (c) diagonal to different people, what were their reactions and how did you feel?

	Seated directly opposite	Adjacent to	Diagonal to
Reactions Received			
Feeling of Self			

34 A FIELD TRIP

LESSON OBJECTIVES

This field trip is designed to accomplish the following objectives:

1. Acquaint the student with a local facility of scientific interest.

2. Allow the student to express his or her impressions relating to the field trip experience.

3. Acquaint the student with some areas of scientific study outside of the classical biological field.

MATERIALS NEEDED

1. Laboratory reports sheets for this exercise
2. Pen or pencil

LABORATORY EXERCISE PROCEDURE

The instructor will provide information as to which area(s) this field trip covers and may make specific assignments. Consult the instructor for this information keeping in mind any time constraints such as the due date for the laboratory report.

While proceeding through the exhibits(s), consult the lab report and complete the sections. When the lab report is complete—*it must be legible*—submit it to the instructor for credit.

LABORATORY EXERCISES

Space is provided to discuss three major exhibits, experiments, or activities in the chosen facility. Complete each section.

A FIELD TRIP

Report Sheet 1

Name _____

Student ID # _____

Campus _____

Date _____

I.   A. Name of Exhibit _____

     B. List the important physical items in this exhibit. (Models, machines, etc.)

_____

_____

_____

     C. Explain the significance of this exhibit. _____

_____

_____

_____

     D. Make suggestions for improvement of this exhibit. _____

_____

_____

_____

II. A. Name of Exhibit _____

     B. List the important physical items in this exhibit. (Models, machines, etc.)

_____

_____

_____

     C. Explain the significance of this exhibit. _____

_____

_____

_____

A Field Trip
Report Sheet 2

Student ID # _____

    D.   Make suggestions for improvement of this exhibit. _____

_____

_____

_____

III. A.   Name of Exhibit_____

    B.   List the important physical items in this exhibit. (Models, machines, etc.)

_____

_____

_____

    C.   Explain the significance of this exhibit. _____

_____

_____

_____

    D.   Make suggestions for improvement of this exhibit. _____

_____

_____

_____

A Field Trip
Report Sheet 3

Student ID # _____

Additional Notes or Comments Page